National Vegetatio
Field guide tc

Cover photograph:
Bovey Valley Woodlands,
National Nature Reserve, Devon.
© English Nature

National Vegetation Classification: Field guide to woodland

J.E. Hall, K.J. Kirby and A.M. Whitbread

Joint Nature Conservation Committee
Monkstone House
City Road
Peterborough
PE1 1JY
UK

ISBN 1 86107 554 5

© JNCC 2004

First edition 2001
Revised reprint 2004

Contents

Preface		7
1.	**Introduction**	9
	National Vegetation Classification	9
	Woodland section of the NVC	9
	Key to the woodland section of the NVC	10
2.	**Key to woodlands and scrub**	13
3.	**Community descriptions and sub-communities keys**	25
	W1 *Salix cinerea – Galium palustre* woodland	26
	W2 *Salix cinerea – Betula pubescens – Phragmites australis* woodland	28
	W3 *Salix pentandra – Carex rostrata* woodland	32
	W4 *Betula pubescens – Molinia caerulea* woodland	34
	W5 *Alnus glutinosa – Carex paniculata* woodland	38
	W6 *Alnus glutinosa – Urtica dioica* woodland	42
	W7 *Alnus glutinosa – Fraxinus excelsior – Lysimachia nemorum* woodland	46
	W8 *Fraxinus excelsior – Acer campestre – Mercurialis perennis* woodland	50
	W9 *Fraxinus excelsior – Sorbus aucuparia – Mercurialis perennis* woodland	57
	W10 *Quercus robur – Pteridium aquilinum – Rubus fruticosus* woodland	61
	W11 *Quercus petraea – Betula pubescens – Oxalis acetosella* woodland	67
	W12 *Fagus sylvatica – Mercurialis perennis* woodland	72
	W13 *Taxus baccata* woodland	76
	W14 *Fagus sylvatica – Rubus fruticosus* woodland	78
	W15 *Fagus sylvatica – Deschampsia flexuosa* woodland	81
	W16 *Quercus* spp. – *Betula* spp. – *Deschampsia flexuosa* woodland	84
	W17 *Quercus petraea – Betula pubescens – Dicranum majus* woodland	89
	W18 *Pinus sylvestris – Hylocomium splendens* woodland	94

4.	References		98
5.	Further reading		99
Appendices			
	I	Relationships between different woodland classification systems	100
	II	Floristic tables	103
	III	Key bryophytes	108
	IV	The distribution of NVC data available for woodlands	112
	V	Latin–English list of tree and shrub species	114
	VI	A minimalist approach to data collection for use with the NVC key	115

Preface

The woodland section of the National Vegetation Classification (NVC) was published in 1991 (Rodwell 1991). Since then, English Nature and the Joint Nature Conservation Committee have produced a variety of material to promote and assist in its use (Kirby, Saunders and Whitbread 1991; Palmer 1992; Whitbread and Kirby 1992; Cooke and Kirby 1994; Hall 1996, 1997). We have also run a large number of training courses, to introduce people to the use of the woodland NVC.

This volume brings together some of this published and unpublished material and experience in what we hope will prove a useful guide that people can use in the field.

Keith Kirby
Jeanette Hall

Notes on nomenclature

Scientific names of higher plant species are as used in Stace (1997). English names of higher plants are from Dony *et al.* (1986). Scientific names of lower plants are as used in Blockeel & Long (1998) and Coppins (2002). The names of National Vegetation Classification (NVC) types are from Rodwell (1992). In some cases these contain species names which differ from those used in Stace (1997); where such inconsistencies occur, the names used in Rodwell (1992) have been retained. Both current species names and those used in Rodwell (1992) are given in Apendices II and III.

1 Introduction

National Vegetation Classification

Since its development in the 1980s, the NVC has become the standard classification used for describing vegetation in Britain. Whereas many other classifications are restricted to particular types of vegetation (e.g. the Stand Type classification which describes only woodland (Peterken 1981)), the NVC aims to describe all the vegetation of Great Britain. This means that it is possible to analyse, and map, a complex site, composed of several habitat types (e.g. woodland, scrub, heathland and bog) using the same classification system. Successional or treatment related changes in the vegetation, for example between open glades, shaded rides and the vegetation of clear-fells can be more easily described than is possible using other classifications.

The NVC is a 'phytosociological' classification, classifying vegetation solely on the basis of the plant species of which it is composed. The resulting communities can usually be correlated to other factors, especially geology and soils, age and management; but the plant species alone are used to assign the vegetation to a community.

The NVC breaks down each broad vegetation type (e.g. woodland, calcareous grassland, mires) into communities, designated by a number and name (e.g. W8 *Fraxinus excelsior – Acer campestre – Mercurialis perennis* woodland, CG1 *Festuca ovina – Carlina vulgaris* grassland, M10 *Carex dioica – Pinguicula vulgaris* mire). Many (but not all) of these communities contain several sub-communities, designated by a letter (e.g. W8a *Fraxinus excelsior – Acer campestre – Mercurialis perennis* woodland *Primula vulgaris – Glechoma hederacea* sub-community). Sub-communities may be further divided into variants (e.g. M10bi and ii) but this has not been adopted within the woodland section of the classification.

Woodland section of the NVC

The NVC woodland classification is based on 2,648 samples from ancient and recent woods throughout Britain (Rodwell 1991). This is the biggest data set yet analysed for the production of a woodland classification in Britain (the Stand Type system, for example, was based on about 800 samples (Peterken 1981)). Apart from the sheer numbers of samples, the geographic and ecological spread of sampling makes it the classification most representative of the range of British woodland. The relationships between the NVC and other woodland classifications are shown in Appendix I.

There are 18 main woodland types and seven scrubs or underscrubs, most of which are divided further to give a total of 73 sub-communities.

Factors other than plant composition are also important in nature conservation terms. Two woods may be of the same vegetation type, but if one is regularly coppiced and the other is high forest the bird and invertebrate life will be very different. Ancient examples of a type are likely to contain more of the species typical of ancient woodlands (e.g. oxlip *Primula elatior*, herb-Paris *Paris quadrifolia*) than recent examples of the same type. The NVC should not therefore be seen as the only way of describing woodland, but rather as one element in such descriptions.

Subsequent to the publication of *British Plant Communities*, various gaps in coverage of the NVC have been identified at community and sub-community level, including several woodland and scrub types (Rodwell *et al.* 2003; Goldberg 2003). No attempt has been made to incorporate these into the present guide, pending further analysis and formal description (Strachan & Jackson 2003).

A seminar was held in 2001 by JNCC and the British Ecological Society to review ten years experience of using the NVC classification for woodlands (Goldberg 2003). Topics covered included the wide range of uses, as well as limitations, of the current classification, consideration of possible future developments, and a European perspective on British woodlands. A phytosociological conspectus in Volume 5 of *British Plant Communities* (Rodwell 2000) also places all NVC communities within a hierarchical framework of European vegetation and gives helpful insight into the floristic relationships of NVC woodland and scrub types.

Key to the woodland section of the NVC

The key presented here can be used in the field or back in the office, using species lists or constancy tables that show the frequency of each species found in a group of samples (see Appendix VI for a minimalist recording protocol). Appendix II provides an explanation of constancy in relation to the species tables. At each stage in the key, two or more possibilities are presented. It is important to read all of these before choosing where to go next. Alternative pathways may need to be considered, particularly if the data are imperfect: for instance, important species like wood anemone *Anemone nemorosa* may have been missed because the wood was surveyed late in the year, or bryophytes may have been ignored because the surveyor could not identify them. For some woodland types, identification of certain bryophytes is important if the community or sub-community is to be correctly allocated. A list of the most common bryophytes used in the classification is given in Appendix III.

Before accepting the result, the composition of the stand should be checked against the floristic tables (see Appendix II) and description for the type. If the stand seems very different to the data in the tables or description, review the sequence of steps that you have taken and see whether an alternative would be a better fit. No stand will be a perfect fit and the following points should be borne in mind by those starting to use the NVC.

1) *Most of the species in a table do not occur in any given stand.* The tables are the summarised results from a wide range of samples throughout the country. In any one stand many of these species will not occur. Conversely species may be recorded in individual stands which do not occur in the summary tables.

2) *One or more of the constant species, including those used to name the community, may be absent.* Constant species are those recorded in 61% or more samples – they were not in all samples. In fact, if four or more constants are specified for a type, one may be absent simply because of chance sampling effects. More intensive survey may find the missing species, but it may just not be present in the stand.

3) *Monocultures of species may occur in the field layer, which are very distinct but difficult to assign beyond community level.* Certain very gregarious species, e.g. dog's mercury *Mercurialis perennis* may occur with virtually no other accompanying species. Stands of hornbeam may have almost no field layer at all. These examples are likely to be W8 and W10 respectively, but it may not be possible to fit them into any sub-community. Great wood-rush *Luzula sylvatica* is another species that may occur as almost a single-species ground-flora under a range of different canopies. Some of these assemblages are being considered for future separation as distinct sub-communities.

4) *Variations in the tree and shrub composition, often caused by forestry treatments, can impart a distinctive appearance and character to many stands without altering the NVC type.* This is particularly the case in lowland Britain, with stands of small-leaved lime or hornbeam. It may also be the case where a wood has been underplanted with beech or non-native conifer species, or where a particular species (e.g oak) has been favoured by foresters. In these circumstances the field layer is often a better guide to the NVC type than the woody layers. This does not

however mean that the composition of the woody layer is unimportant, and these variations in it should be recorded.

5) *The appearance of an area (and sometimes the NVC type) may change, at least temporarily, following felling.* Species-richness increases dramatically, and previously open herb-dominated communities become very grassy. On sites which have not previously been subject to large-scale fellings, some of the changes in species composition will be permanent. On other sites they may be part of a cyclical pattern. During this open stage some stands will be closer to grassland or scrub types than to the parent closed canopy community. NVC communities describe the vegetation as it is and at the open phase the vegetation is often like a rough field.

6) *Differences in grazing levels lead to changes in the appearance of types.* At low levels of grazing in W11, W16 and W17, species such as bilberry *Vaccinium myrtillus* and *Luzula sylvatica* are likely to be prominent, whereas high levels of grazing favours some grasses and bryophytes. These shifts in relative abundance may not affect the classification, but where grazing differences have been maintained for many years the boundaries between sub-communities may be determined by these grazing patterns. Recently, grazing by deer in lowland woods has become an issue as well. Grasses such as false brome *Brachypodium sylvaticum* and tufted hair-grass *Deschampsia cespitosa* have spread through W8 type woodland, blurring the sub-community differences. Bramble *Rubus fruticosus* has become much less abundant, so the appearance of many other woods has also changed.

7) *Not all samples, however carefully collected, can be matched to just one set of summary tables.* The NVC types are a set of defined points in the continuum of woodland variation. Intermediate stands, e.g. between W10 and W11, do occur. Our experience is that most stands can be identified as closer to one type than to another, but rarely is the fit perfect.

8) *A type may be identified in a place not shown on the published distribution maps.* The maps published in this report only show where data was collected. There are gaps in the availability of data, particularly in the English midlands. Appendix IV shows an updated distribution of all woodland NVC records. They give a good indication of the range of the type but are not definitive. As further information is collected a clearer picture will emerge. The most recent published distribution maps are available in Hall (1997). In order to improve our knowledge of the distribution of NVC types please send additional woodland records, particularly those with supporting quadrat data, to Keith Kirby, English Nature, Northminster House, Peterborough PE1 1UA.

2 Key to woodlands and scrub

The community key is designed to enable the user to identify stands to the NVC community level. Read through all the alternatives before picking the one that fits best. Additional guidance is given in text boxes to aid the separation of close communities, or where geographical variation within communities can lead to confusion. Sub-community keys and short descriptions follow the community key, but these are only summaries. Any conclusions should be checked at least periodically against the full published floristic tables and descriptions.

English names are used for the more common trees and shrubs, and scientific names for field layer and bryophyte layer species. A Latin-English species list for trees and shrubs is given in Appendix V. Each step of the key is numbered, and the number of the previous step is given in brackets, e.g. 20 (1) = Step 20, from Step 1.

1   The first step separates out scrub types from woodland proper. In terms of structure and composition, woodland and scrub grade into each other, so it is difficult to devise a definitive boundary between them. If in doubt, work through the whole key starting at 2.

(a) Low scrub with *Salix lapponum*, and sometimes *S. lanata*, *S. myrsinites* or *S. reticulata*, with luxuriant mixtures of *Vaccinium myrtillus*, *V. vitis-idaea* and *Empetrum nigrum* ssp. *hermaphroditum*, *Luzula sylvatica*, *Deschampsia cespitosa*, tall dicotyledons and bryophytes.

Habitat: A rare community of montane crags and ledges.

W20 *Salix lapponum – Luzula sylvatica* scrub

OR   **(b)** Scrub or underscrub dominated by one or more of hawthorn, juniper, blackthorn, elder, *Ulex europaeus*, *Cytisus scoparius*, *Rosa canina* agg. or *Rubus fruticosus* agg. Trees and saplings may be numerous but don't form an overtopping canopy. → 20

OR   **(c)** High forest or coppice with a proper (sometimes quite open) canopy. The above shrubs can be frequent, but not dominant, often forming an understorey with other species. → 2

2   This second division separates off 'wet' woodlands from those with normally drier field layers. Stands of birch on dry

ground, e.g. regeneration on former open ground or following felling, should normally be put through the drier ground option (step **10**).

(a) Canopy dominated by one or more of alder, willow or birch.

Habitat: Wet or poorly drained ground. ➜ 3

OR **(b)** Canopy dominated by other species (alder, beech, yew, pine, oak, ash, etc). Willow and birch may be present but usually in low quantities.

Habitat: Free-draining to poorly-drained sites, but if the latter then usually mineral soils (often heavy clay) rather than organic soils. ➜ 10

Wet Woodland

North-western stands of all wet woodland types may contain *Salix aurita* instead of *S. cinerea*.

3 (2) **(a)** Dominated by *Salix cinerea* (*S. aurita*) and/or downy birch. *Salix pentandra* may be present but other woody species are usually infrequent. ➜ 4

OR **(b)** Dominated by alder and/or *Salix fragilis* ➜ 7

OR **(c)** Dominated by one or more of *Salix purpurea*, *S. triandra*, *S. viminalis* or hybrids, forming scrubby vegetation or osier beds.
W6c *Alnus glutinosa* – *Urtica dioica* woodland, *Salix viminalis*/*triandra* sub-community

4 (3) **(a)** Single tree/shrub layer composed of mixtures of *Salix cinerea* (*S. aurita*) and *S. pentandra* with occasional downy birch.

Swampy field layer with abundant *Carex rostrata*, *Equisetum fluviatile* or *Menyanthes trifoliata*. Five or more of *Angelica sylvestris*, *Caltha palustris*, *Cardamine pratensis*, *Crepis paludosa*, *Filipendula ulmaria*, *Galium palustre*, *Geum rivale*, *Valeriana dioica*. Bryophyte mat often extensive with *Calliergonella cuspidata*, *Climacium dendroides*, *Eurhynchium praelongum*, *Mnium hornum*, *Rhizomnium punctatum* and occasional patches of *Sphagnum palustre*, *S. fallax* and/or *S. squarrosum*.
W3 *Salix pentandra* – *Carex rostrata* woodland

OR	**(b)** *Salix pentandra*, and/or the field layer species and bryophytes listed above, absent. → 5
5 (4)	**(a)** Tree/shrub layer with frequent or abundant downy birch and *Salix cinerea* (*S. aurita*). Field layer with frequent and often abundant *Phragmites australis*, but *Carex paniculata*, *Lythrum salicaria* and *Lysimachia vulgaris* generally infrequent. **W2 *Salix cinerea* – *Betula pubescens* – *Phragmites australis* woodland**
OR	**(b)** Tree/shrub layer with either downy birch or *Salix cinerea* (*S. aurita*) markedly more frequent and abundant than the other. Field layer with rare or absent *Phragmites australis*. → 6
6 (5)	**(a)** Tree/shrub layer with frequent and generally abundant *Salix cinerea* (*S. aurita*) and occasional downy birch. Field layer somewhat varied. Usually frequent *Galium palustre* and *Mentha aquatica*, but *Molinia caerulea* is rare or absent. **W1 *Salix cinerea* – *Galium palustre* woodland**
OR	**(b)** Canopy usually well-defined, although often quite open and somewhat moribund, with frequent and generally abundant downy birch and occasional *Salix cinerea* (*S. aurita*). Field layer with constant and often abundant *Molinia caerulea*. *Sphagnum* spp. and/or *Polytrichum commune* patches may also be common. **W4 *Betula pubescens* – *Molinia caerulea* woodland**

> Alder may be present at a low frequency in W4 and some of the sub-communities are similar to those of W6 *Alnus glutinosa* – *Urtica dioica* woodland. If in doubt check both sets of descriptions.

7 (3)	**(a)** Field layer with frequent and generally abundant *Carex paniculata* or, locally, *Scirpus sylvaticus*, with some (not necessarily all) of *Carex acutiformis*, *Cirsium palustre*, *Dryopteris dilatata*, *Eupatorium cannabinum*, *Filipendula*

| | Field layer with *Hedera helix*, *Mercurialis perennis* or other species (e.g. *Allium ursinum*, *Brachypodium sylvaticum*, *Circaea lutetiana* or *Sanicula europaea*) of base-rich rather than mesotrophic soils. The field layer may be sparse if yew is abundant. |

W12 *Fagus sylvatica* – *Mercurialis perennis* woodland

OR **(b)** Canopy and Shrub layer usually with two or more of silver birch, holly, sessile oak, pedunculate oak, beech saplings or silver birch saplings.

Field layer with *Deschampsia flexuosa*, *Vaccinium myrtillus* or *Calluna vulgaris*. May be sparse if beech canopy is very dense.

W15 *Fagus sylvatica* – *Deschampsia flexuosa* woodland

Yew woodland

13 (10) **(a)** Canopy dominated by yew.

W13 *Taxus baccata* woodland

OR **(b)** Canopy dominated by other species (oak, ash, lime, elm, pine, etc). Yew absent or, if it is present, then only as rare or scattered trees, and usually in the understorey. → **14**

Pine woodland

14 (13) **(a)** Canopy dominated by Scots pine.

Field layer with two or more of *Calluna vulgaris*, *Deschampsia flexuosa*, *Vaccinium myrtillus*, *V. vitis-idaea*.

Bryophyte layer well-developed with *Dicranum scoparium*, *Hylocomium splendens* and *Pleurozium schreberi* and two or more of *Hypnum jutlandicum*, *Lophocolea bidentata*, *Plagiothecium undulatum*, *Scleropodium purum*, *Ptilium crista-castrensis*, *Rhytidiadelphus loreus*, *R. triquetrus*.

W18 *Pinus sylvestris* – *Hylocomium splendens* woodland

OR **(b)** Canopy dominated by other species (oak, ash, elm, lime, etc). Scots pine absent or, if it is present, then without the above field layer species or bryophytes. → **15**

Strictly speaking, W18 is restricted to native pinewoods and mature pine plantations within the native range of pine, although long-established plantations elsewhere may have similar field and bryophyte layers. In general, plantations of pine in southern Britain are likely to be derived from one of the other woodland communities and should be classified accordingly. For example, self-sown stands on the southern heath will often be closer floristically to acid oak-dominated woodland (usually W16) than to W18.

Oak-dominated and mixed deciduous woodland

The remaining woodland types comprise the 'mixed deciduous' and 'oak' woodland communities, which can be dominated by ash, birch, elm, field maple, hazel, hornbeam, lime or oak. The field layer and, to a lesser extent, the shrub layer, are more useful for distinguishing the different communities than the canopy species. Plantations of non-native species derived from these types may still fit into the communities if a reasonable field layer survives. The key separates off the two most acid communities first (steps 15, 16), then the two types found on mesotrophic soils (steps 17,18), leaving the two on base-rich soils until last (step 19). In each case one of the pair is more common in the north and west and the other in the south and east, but there is considerable overlap and the classification should be made on the botanical composition of the actual samples.

15 (14)   **(a)** Canopy usually a mixture of oak (usually sessile oak) and birch (usually downy birch).

Shrub layer can include hazel, holly and rowan, but is often sparse.

Field layer usually contains *Deschampsia flexuosa*, *Oxalis acetosella*, *Pteridium aquilinum* and *Vaccinium myrtillus*. May be very sparse.

Bryophyte layer well-developed with six or more of *Dicranum majus, Dicranum scoparium, Hylocomium splendens, Plagiothecium undulatum, Pleurozium schreberi, Polytrichum formosum, Rhytidiadelphus loreus, Thuidium tamariscinum.*

W17 ***Quercus petraea* – *Betula pubescens* – *Dicranum majus* woodland**

> In western woods, the characteristic bryophytes of W17 may be found on rocks within a generally more bryophyte-poor, grass-dominated type. Such mosaics may be assigned to one or the other type according to which is more abundant, or may be recorded as an intimate mosaic. In ungrazed woods the bryophyte layer may be much less abundant, and growing under *Vaccinium*, but the combination of species is usually still present.

OR **(b)** Bryophyte layer lacking this combination of species although oak, birch and the field layer species may be present.
➜ 16

16 (15) **(a)** Canopy usually a mixture of oak and birch species, or self-sown pine stands on heaths, or plantations of pine, larch or Douglas fir on acid soils.

Shrub layer usually includes rowan. May be very sparse.

Field layer species-poor with *Deschampsia flexuosa* and *Pteridium aquilinum*, and *Vaccinium myrtillus* in ungrazed woods. 'Western' bryophytes absent. *Oxalis acetosella* and grasses such as *Agrostis capillaris*, *Anthoxanthum odoratum* and *Holcus mollis* are rare.

W16 *Quercus* spp. – *Betula* spp. – *Deschampsia flexuosa* woodland

OR **(b)** Canopy with oak and birch, or a mixture of other species. Field layer richer or more grass-dominated. ➜ 17

17 (16) **(a)** Canopy usually a mixture of oak (usually sessile oak or a mixture of sessile and pedunculate oak) and birch.

Shrub layer often includes rowan and hazel.

Field layer frequently dominated by grasses, with six or more of *Agrostis capillaris*, *Anthoxanthum odoratum*, *Deschampsia flexuosa*, *Galium saxatile*, *Holcus mollis*, *Oxalis acetosella*, *Potentilla erecta*, *Pteridium aquilinum*, *Viola riviniana*.

Bryophyte layer can be extensive with two or more of *Hylocomium splendens*, *Scleropodium purum*, *Rhytidiadelphus squarrosus* and *Thuidium tamariscinum*.

W11 *Quercus petraea* – *Betula pubescens* – *Oxalis acetosella* woodland

> Mosaics of W11 and W17 may occur where bryophyte-rich boulders are interspersed with deeper grass-dominated hollows (see above). On more calcareous and base-rich substrates to the west and north, W11 may grade into communities dominated by ash, elm, hazel or sycamore, where *Mercurialis perennis* and other calcicolous herbs or grasses are common. This is particularly likely around streams, at the base of slopes, etc. Oak woodland without the grass – herb field layer typical of W11 should not be assigned to this community just because it is dominated by sessile oak.

OR | **(b)** Canopy with oak (usually pedunculate oak) and birch, or a mixture of other species.

Field layer without the above combinations of herbs and bryophytes. → 18

18 (17) **(a)** Canopy usually dominated by oak (usually pedunculate oak) and birch, although hornbeam, sweet chestnut and lime may be locally abundant. Ash, elm and sycamore are generally infrequent, but can occur with the field layer typical of this community, especially in the north and west. Plantations of non-native species may fit into this community.

Shrub layer frequently contains hazel and hawthorn.

Field layer usually contains some combination of abundant *Rubus fruticosus* and/or *Pteridium aquilinum* and/or *Lonicera periclymenum*, often with *Hyacinthoides non-scripta* as a vernal dominant. *Mercurialis perennis* and other calcicolous herbs are rare.

W10 ***Quercus robur* – *Pteridium aquilinum* – *Rubus fruticosus* woodland**

OR **(b)** Canopy usually with abundant ash, elm or field maple. Hornbeam or lime may be locally abundant, in which case other trees and shrubs may be scarce. Oak and birch may be present but are not usually common.

Field layer dominated by *Mercurialis perennis* or other calcicolous herbs and grasses (such as *Brachypodium sylvaticum*). → 19

> As with W11 (*Quercus petraea – Betula pubescens – Oxalis acetosella* woodland), W10 may grade into communities dominated by ash, elm or field maple, where *Mercurialis perennis* and other calcicolous herbs or grasses are common. This is particularly likely where localised flushing or base-enrichment occurs. In some areas, particularly around the upland/lowland boundary (e.g. in the Peak District or Welsh borders) it may be difficult to separate W10 and W11. The most useful feature for distinguishing the two is the abundance of *Agrostis capillaris*, *Anthoxanthum odoratum* and *Deschampsia flexuosa* in W11.
>
> Note that lime stands (and, to a lesser extent, hornbeam stands) may be classified either as W10 or as W8 (the equivalent community of base-rich soils) according to their ground flora.

19 (18)   **(a)** Canopy and shrub layer with some of the southern calcicolous shrubs (field maple, dogwood, Midland hawthorn, spindle, wayfaring tree) and/or hornbeam, suckering elms, or lime. Samples from the north and west often have fewer of these species and sycamore, sessile oak (rather than pedunculate oak) and wych elm may be more common.

Field layer with some of *Allium ursinum*, *Anemone nemorosa*, *Brachypodium sylvaticum*, *Deschampsia cespitosa*, *Filipendula ulmaria*, *Geranium robertianum*, *Glechoma hederacea*, *Hedera helix*, *Hyacinthoides non-scripta*, *Mercurialis perennis*, *Primula* spp. (including *P. elatior* in eastern England), *Teucrium scorodonia* or *Urtica dioica*. Ferns, especially *Athyrium filix-femina* and *Dryopteris* spp., are often sparse although *Phyllitis scolopendrium* and *Polystichum setiferum* may be common in the west, and the bryophyte layer, although sometimes locally extensive, is usually species-poor.

W8 *Fraxinus excelsior – Acer campestre – Mercurialis perennis* woodland

OR   **(b)** Canopy dominated by ash or, locally, sycamore, sessile oak, wych elm or (in north-west Scotland) hazel and birch. Rowan often scattered through the stand. Southern shrubs usually rare, although lime may be present as isolated trees in northern England and Wales.

Field layer variable. The calcicolous herbs and grasses listed above for W8 are common, as are bryophytes, *Athyrium filix-femina*, *Dryopteris* spp., *Oxalis acetosella* and *Viola riviniana*.

W9 *Fraxinus excelsior* – *Sorbus aucuparia* – *Mercurialis perennis* woodland

As for W10 and W11 (see above) it can be difficult to distinguish W8 (mainly south-eastern) and W9 (mainly north-western) at community level, particularly around the upland/lowland boundary (e.g. in the Peak District or Welsh borders). *Arum maculatum* and the southern shrubs indicate a tendency towards W8, whereas *Oxalis acetosella* and rowan are more typical of W9. It may be easier to consider the sub-communities of both W8 and W9 together to see which fits best.

Scrub communities

20 (1)	**(a)** Low scrub (usually <2 m) dominated by *Rubus fruticosus* agg. or *Pteridium aquilinum* with few other woody species. → 21
OR	**(b)** Taller vegetation (up to 5 m or more high) dominated by one or more of hawthorn, juniper, blackthorn, *Ulex europaeus* and *Cytisus scoparius*, although *Rubus fruticosus* agg. is often a prominent component of the vegetation. → 22
21 (20)	**(a)** *Pteridium aquilinum* abundant, often with *Rubus fruticosus* agg. **W25 *Rubus fruticosus* agg. – *Pteridium aquilinum* underscrub**
OR	**(b)** *Pteridium aquilinum* absent. **W24 *Rubus fruticosus* agg. – *Holcus lanatus* underscrub**
22 (20)	**(a)** Canopy dominated by juniper. Often open, occasionally with overtopping downy birch. Field layer with frequent *Agrostis capillaris*, *A. vinealis*, *Galium saxatile*, *Luzula pilosa*, *Oxalis acetosella* and *Vaccinium myrtillus*. Bryophyte layer well-developed with *Hylocomium splendens* and *Thuidium tamariscinum*.

	W19 *Juniperus communis – Oxalis acetosella* **woodland**
OR	**(b)** Juniper absent or, if it is present, then not with the above field layer species. → 23

> Juniper stands on the southern chalk are better referred to W21d. Juniper also occurs as an understorey in pine stands and less frequently birch or oak stands. Normally it should be relatively easy to decide whether a stand is a scrub community with only occasional trees (W19) or a juniper understorey in what may be locally rather open pine, birch or oak woodland. The floristic differences can, however, be small.

23 (22)	**(a)** Dominated by *Ulex europaeus* and/or *Cytisus scoparius*, with *Rubus fruticosus* agg. but few other woody species.
	W23 *Ulex europaeus – Rubus fruticosus* **agg. scrub**
OR	**(b)** Dominated by other woody species. *Ulex europaeus* and *Cytisus scoparius* occasionally present as a minor component. → 24
24 (23)	**(a)** Dominated, often solely, by blackthorn with few other woody plants.
	W22 *Prunus spinosa – Pteridium aquilinum* **scrub**
OR	**(b)** Dominated by other woody species, especially hawthorn. Although blackthorn may be frequent, and sometimes locally prominent, other species usually exceed it in cover.
	W21 *Crataegus monogyna – Hedera helix* **scrub**

3 Summary community descriptions and sub-communities keys

The summary descriptions provided here are derived from the full accounts prepared by John Rodwell. Only the true woodland communities (W1 to W18) are dealt with here. The entry for each community is presented as follows:

- brief description of community
- map of the distribution of the community in Great Britain[1]
- key to sub-communities (where described)
- brief description of each sub-community

These descriptions are not intended as a substitute for the full accounts provided in Volume 1 of *British Plant Communities* (Rodwell 1991) but as an *aide memoire* to assist surveyors in the field. Anyone who uses this book should always check their results against the frequency tables and full descriptions for each community in that volume. The caveats made regarding the use of the community key should also be borne in mind when using the keys to sub-communities.

Stands may be recorded which fit well to a community, but do not contain any of the preferentials required to assign a sub-community. Such stands should be recorded as undifferentiated examples of the community. Other stands may be found with preferentials from two or more sub-communities. Such stands will usually tend more towards one or the other, but in a natural continuum some stands will occasionally occur which cannot be objectively allocated to one or other sub-community. These should be recorded as an intermediate stand (e.g. W8a/W8b). The use of the 'intermediate' category should however be avoided as far as possible. There is often a tendency, particularly in the learning stage, to regard most stands as intermediate because a very high level of fit is expected.

---

1 There are no records for any woodland type in Orkney or Shetland in the NVC Woodland Database and so these islands have not been included on the distribution maps.

W1 *Salix cinerea – Galium palustre* woodland

A community of wet mineral soils on the margins of standing or slow-moving water and in moist hollows, mainly in the lowlands. It often occurs as a narrow fringe or as scattered fragments around ponds, lakes, dune slacks, etc.

The canopy is dominated by *Salix cinerea* but its structure is irregular. Young stands often consist of a mass of bushes of variable height, older stands are more regular with a single tier of sallows c. 4-8 m high. Other woody associates are only occasional – birch with scarce alder, pedunculate oak and silver birch. Other *Salix* spp. are uncommon but there can be scattered hawthorn, hazel and alder buckthorn.

The field layer varies in its cover and composition but the general appearance is of an open scatter of herbs. *Galium palustre* is common. *Mentha aquatica* and *Juncus effusus* are also frequent with scattered *Angelica sylvestris*, *Lycopus europaeus*, *Ranunculus flammula*, *R. repens*, *Epilobium palustre*, *Equisetum fluviatile*, *Filipendula ulmaria*, *Cirsium palustre*, *Rumex sanguineus*, *Caltha palustris*, *Hydrocotyle vulgaris*, *Potentilla palustris* and *Iris pseudacorus*. Scramblers such as *Rubus fruticosus*, *Solanum dulcamara* and *Hedera helix* may be abundant. In other cases the field layer has a grassy appearance – *Holcus lanatus*, *Agrostis canina* and *A. stolonifera*. Generally swamp and fen dominants are rare but occasional stands have some *Carex paniculata*, *C. riparia*, *C. vesicaria* or *Phragmites australis*. Bare ground or with a patchy cover of bryophytes can be quite extensive. *Eurhynchium praelongum* is most frequent with some *Chiloscyphus pallescens*, *Calliergonella cuspidata*, *C. cordifolium*, *Brachythecium rutabulum* and *Rhytidiadelphus squarrosus*. Epiphytic lichens may be conspicuous in sheltered situations in south-west Britain.

No sub-communities.

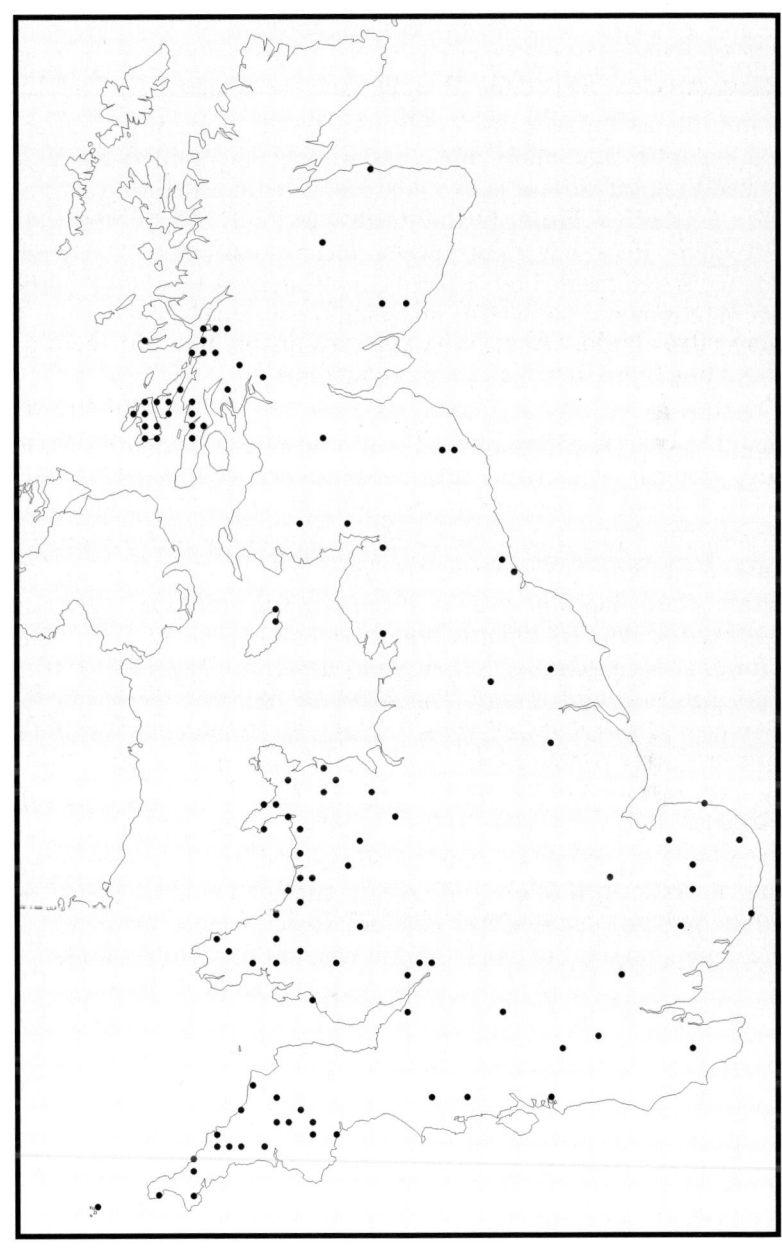

Figure 1 Distribution of W1 *Salix cinerea – Galium palustre* woodland

W2 *Salix cinerea – Betula pubescens – Phragmites australis* woodland

A community of topogenous fen-peats of flood plain mires, terraces of river valley mires and, more rarely, basin mires where litter accumulation has raised the peat surface above the level of winter flooding.

Salix cinerea (*S. aurita* in the north-west) and downy birch are the most frequent trees, but alder may be locally abundant. Their relative abundance is determined by order of colonisation as much as by differing habitat requirements, so there is no specific sequence of succession of the preceding fen. Other woody species which may be locally dominant, particularly in the early stages of colonisation, include alder buckthorn and buckthorn.

The field layer is derived from the preceding fen communities, which are very variable, so there are few constant species. *Phragmites australis* is usually present, either as dense stands or scattered individuals. Other fen dominants which occur sporadically, especially under young, open canopies, include *Calamagrostis canescens*, *C. epigejos*, *Carex acutiformis* and *Cladium mariscus*. *Carex paniculata* may occur, but is more typical of W5. *Thelypteris palustris* can be frequent, with scattered *Eupatorium cannabinum*, *Filipendula ulmaria*, *Lysimachia vulgaris*, *Lythrum salicaria* and, rarely, *Peucedanum palustre*. Such species are more typical of the rich fen *Alnus – Filipendula* sub-community (W2a). Tangles of *Rubus fruticosus* or *Rosa canina* are often present and, less commonly, *Rubus idaeus*, *Ribes nigrum* and *R. rubrum*. *Dryopteris dilatata*, usually uncommon in fens, may be present. Bare ground, and loose mats of *Poa trivialis* and *Eurhynchium praelongum*, can be extensive.

The floristic differences between the sub-communities reflect variation in base-richness and calcium levels in the peat, which are largely dependent on the height and movement of ground water.

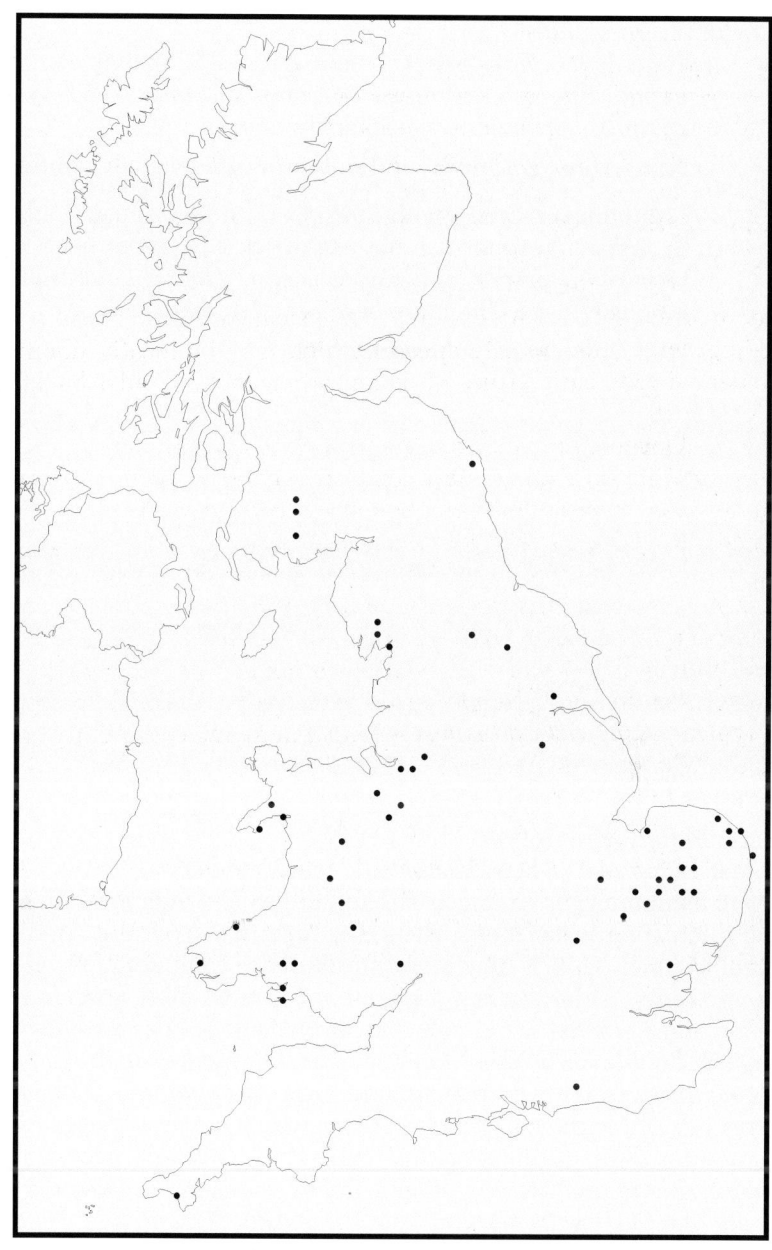

Figure 2 Distribution of W2 *Salix cinerea – Betula pubescens – Phragmites australis* woodland

Key to sub-communities

Field layer with one or more of the following abundant: *Eupatorium cannabinum*, *Filipendula ulmaria*, *Phragmites australis*, *Urtica dioica*. Sphagna rare.

W2a *Alnus glutinosa – Filipendula ulmaria* sub-community

OR

Ground layer always with Sphagna (*S. fimbriatum, S. palustre, S. fallax, S. squarrosum*), which may be prominent. Field layer often 'grassy' with *Agrostis* spp., *Holcus lanatus* and/or *Juncus effusus*. The more base-rich 'fen' herbs are rare.

W2b *Sphagnum* sub-community

Alder may readily invade W2a, and become abundant, so that there is a complete gradation between W2a and W6a (*Alnus glutinosa - Urtica dioica* woodland, typical sub-community). In typical W6a, alder is often overwhelmingly dominant in the canopy. Species such as *Galium aparine* and *Poa trivialis* tend to be more common, whereas the tall sedges and fen herbs, typical of W2a, are less so.

A carpet of Sphagna, as found in W2b, is also characteristic of the *Sphagnum* sub-community of W4 (*Betula pubescens – Molinia caerulea* woodland). W2b tends to have more *Phragmites australis* and *Salix cinerea* and, often, a rather open scrubby structure. *Molinia caerulea* may occur but is much more typical of W4c.

Sub-community descriptions
W2a *Alnus glutinosa – Filipendula ulmaria* sub-community
This sub-community is characteristic of fen peats, which are influenced by the fluctuating water table. The pH is high (6.5–7.5), and conditions fairly eutrophic. This is the more structurally complex, species-rich, sub-community. Alder is often frequent and can be more common than birch or *Salix cinerea*. Ash, oak, hawthorn and guelder rose can also occur. The field layer is often dominated by *Phragmites australis* and/or *Carex acutiformis* with frequent *Eupatorium cannabinum*, *Filipendula ulmaria* and *Urtica dioica*. Other tall herbs which are preferential for this sub-community include *Angelica sylvestris*, *Cirsium palustre* and *Epilobium hirsutum*. There is often a *Rubus fruticosus* underscrub with climbers like *Calystegia sepium*, *Galium aparine*, *G. palustre*, *Humulus lupulus*, *Lonicera periclymenum* and *Solanum dulcamara*. Ferns are less abundant than in W2b and small herbs are not numerous apart from mats of *Poa trivialis*, scattered *Mentha aquatica* and *Caltha palustris*. Patches

of bare earth and peat may have extensive bryophyte carpets but few species are involved, principally *Brachythecium rutabulum, Eurhynchium praelongum* and *Plagiomnium undulatum*. Sphagna are characteristically scarce.

W2b *Sphagnum* sub-community
This sub-community is found where peat levels are isolated from the effects of ground water; either where peat accumulation has raised levels or on floating peat rafts, where the peat level is always above water level. Downy birch is the most abundant woody species with frequent *Salix cinerea*, but alder and ash are less common than in W2a. Alder buckthorn and *Salix aurita* are local but oak, buckthorn and guelder rose are characteristically absent. *Myrica gale* and *Salix repens* can form a patchy lower tier with some *Rubus fruticosus, Rosa canina* and *Lonicera periclymenum*. *Phragmites australis* remains frequent, but *Carex acutiformis* is absent, and other fen monocots and tall fen herbs are sparse. Grasses are often abundant, including *Agrostis canina, Agrostis stolonifera, Holcus lanatus, Molinia caerulea* and *Poa trivialis*. Ferns are also characteristically abundant, including *Athyrium filix-femina, Dryopteris carthusiana, D. cristata, D. dilatata, Thelypteris palustris* and more rarely *Phegopteris connectilis* and *Osmunda regalis*. Sphagna, including *S. fimbriatum, S. squarrosum, S. palustre* and *S. fallax* are generally very abundant, and sometimes form a virtually continuous cover.

W3 *Salix pentandra* – *Carex rostrata* woodland

A community of peat soils kept moist by moderately base-rich and calcareous ground water in open water transitions, most common in northern Britain. Its general geographic limits, particularly in the south, are heavily influenced by climate, many of the species characterising W3 tending to have a northerly distribution.

This type is fairly constant in its composition and structure. The canopy is low, uneven-topped and dominated by *Salix* spp. usually *S. pentandra* and/or *S. cinerea*. Other *Salix* spp. are rare but can be locally abundant – *S. nigricans*, *S. phylicifolia* and *S. aurita*, more rarely *S. viminalis* and *S. purpurea*. Downy birch occurs occasionally but alder is rare. Southern fen species such as alder buckthorn, buckthorn and guelder rose are generally absent.

The field layer can vary widely. Many stands have several species co-dominating, but the overall assemblage of species is distinctive. Tall herbs and horsetails are the most prominent feature, for example *Filipendula ulmaria*, *Angelica sylvestris*, *Valeriana dioica*, *V. officinalis*, *Geum rivale*, *Cirsium palustre* and *Equisetum fluviatile*, but rich-fen species (e.g. *Eupatorium cannabinum*, *Lysimachia vulgaris*, *Lythrum salicaria*, *Iris pseudacorus*) are usually absent. Shorter herbs often form a patchy lower layer, for example *Cardamine pratensis* and *Crepis paludosa* and lesser amounts of *Caltha palustris*, *Mentha aquatica*, *Lychnis flos-cuculi*, *Ranunculus repens*, *Poa trivialis*, *Dactylorhiza fuchsii*, *Equisetum palustre*, *Menyanthes trifoliata*, *Potentilla palustre* and *Galium palustre*. Ferns are not a prominent feature. Large grasses, rushes and sedges may or may not be abundant. *Carex rostrata* occurs most frequently but usually as sparse scattered shoots. Less frequent are *C. diandra*, *C. lasiocarpa*, *C. appropinquata*, *C. paniculata*, *C. laevigata*, *C. vesicaria*, *C. nigra*, *C. curta*, *Juncus acutiflorus* and *J. effusus*. Bryophytes are abundant, sometimes forming a complete ground carpet. *Calliergonella cuspidata*, *Climacium dendroides* and *Rhizomnium punctatum* tend to be the most conspicuous, with some *Plagiomnium affine*, *P. ellipticum*, *P. rostratum*, *P. elatum*, *Mnium hornum* and *Eurhynchium praelongum*. Patches of *Sphagnum* spp. may be locally abundant.

No sub-communities.

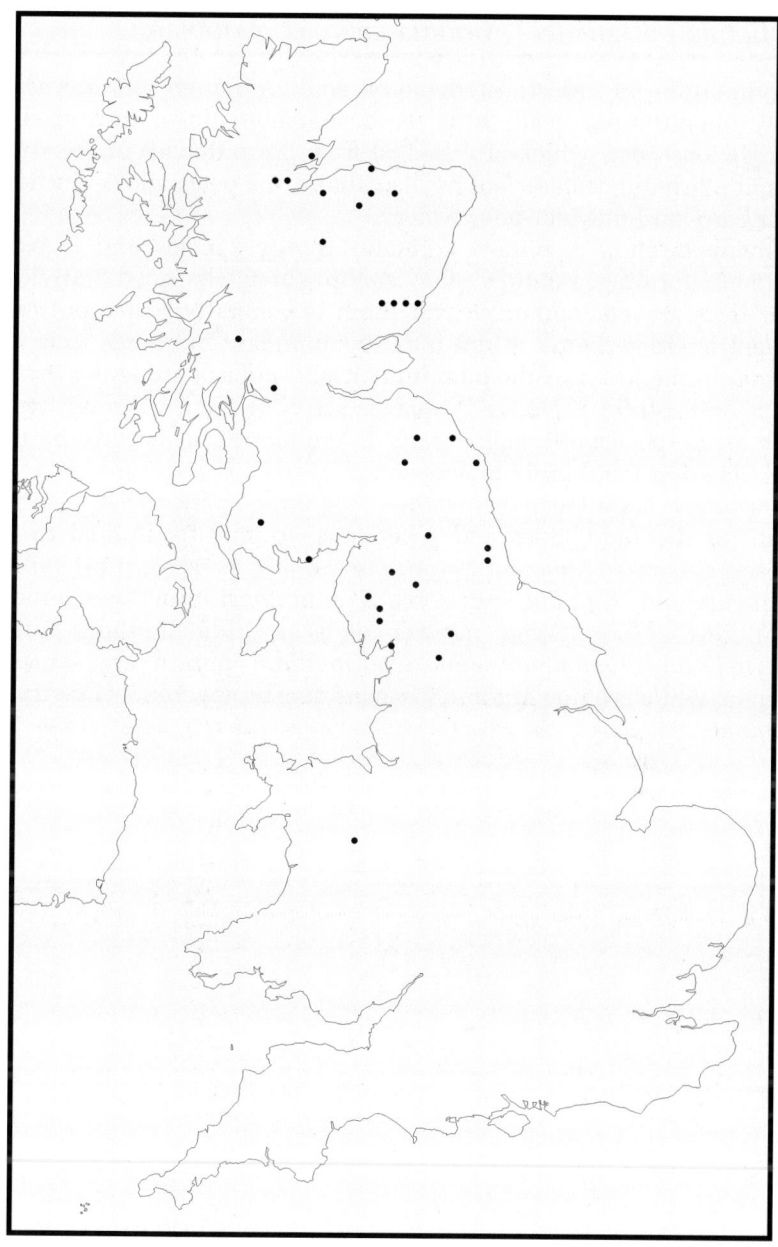

Figure 3 Distribution of W3 *Salix pentandra – Carex rostrata* woodland

W4 *Betula pubescens – Molinia caerulea* woodland

A community of moist, moderately acidic, though not necessarily highly oligotrophic, peaty soils. It is characteristic of thin or drying ombrogenous peats which are isolated from the influence of base-rich or eutrophic ground waters, but is also found on peaty gleys flushed by rather base- and nutrient-poor water.

Downy birch is the most common woody species and is usually dominant, forming a rather open canopy of well-spaced individuals. Other trees are uncommon. Silver birch is generally scarce but can be frequent in drier stands. Alder is rarely abundant but tends to be more frequent in the *Juncus* sub-community. Oaks and ash are very scarce.

The understorey is generally sparse. *Salix cinerea* is the most frequent shrub layer species although locally *S. caprea*, *S. pentandra*, *S. aurita*, hazel, hawthorn and holly may occur.

The great abundance of *Molinia caerulea* is the most distinctive feature of the field layer and other species may be limited to areas between tussocks. *Sphagnum* spp. are usually present, most typically *S. palustre* and *S. fallax* with some *S. subnitens*, sometimes forming a continuous carpet. Other mosses such as *Aulacomnium palustre*, *Eurhynchium praelongum* and *Scleropodium purum* are sometimes common, while eroding *Molinia* tussocks may be covered by *Polytrichum commune*.

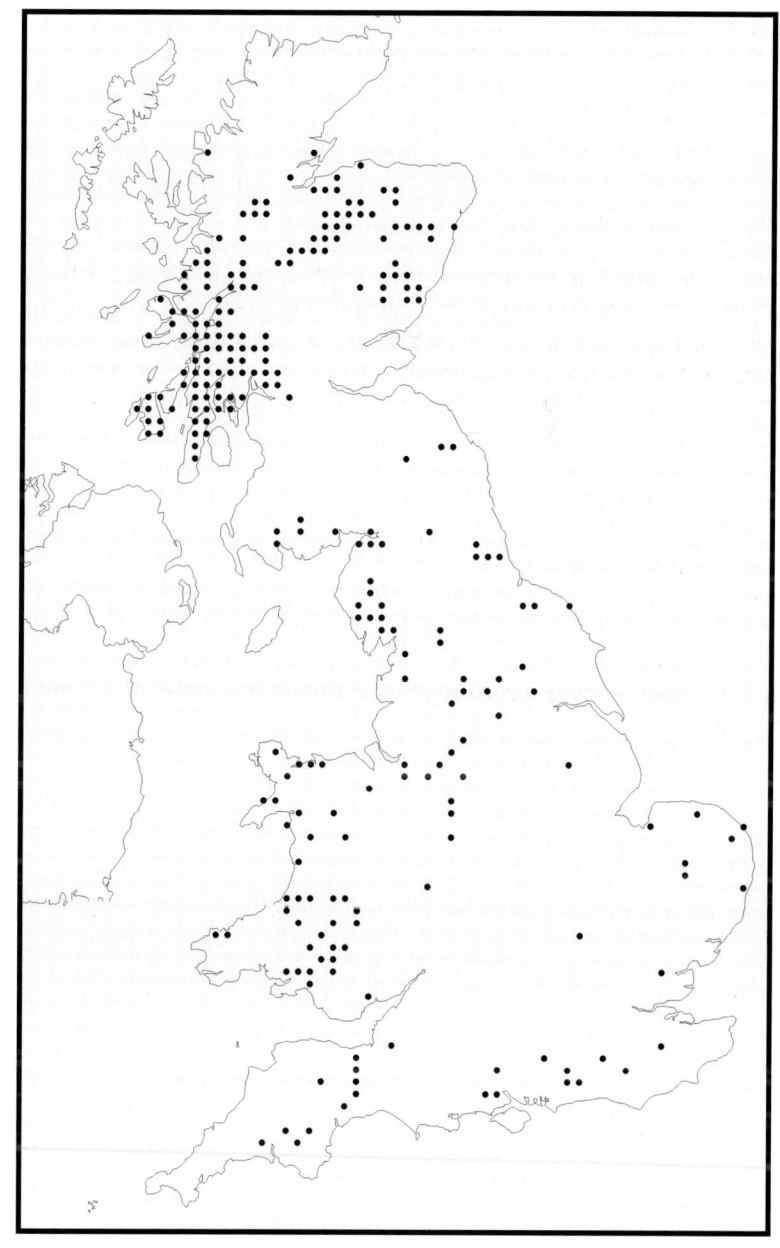

Figure 4 Distribution of W4 *Betula pubescens – Molinia caerulea* woodland

Key to sub-communities

        Ground layer dominated by mixtures of *Sphagnum fimbriatum*, *S. palustre*, *S. papillosum*, *S. fallax* and *S. squarrosum*. *Molinia caerulea* is the only prominent grass. *Deschampsia cespitosa*, *Dryopteris dilatata*, *Holcus* spp., *Lonicera periclymenum*, *Potentilla erecta* and *Rubus fruticosus* are absent or very sparse.

        W4c *Sphagnum* sub-community

OR   Ground layer with frequent *Cirsium palustre*, *Hydrocotyle vulgaris*, *Juncus effusus*, *Potentilla erecta*, *Viola palustris* and other grasses in addition to *Molinia caerulea* (e.g. *Deschampsia cespitosa*, *Holcus mollis* and/or *Holcus lanatus*). *Sphagnum fallax* or, less frequently, *S. palustre* may also be prominent. *Dryopteris dilatata*, *Lonicera periclymenum* and *Rubus fruticosus* are absent or sparse.

        W4b *Juncus effusus* sub-community

OR   Field layer with frequent *Dryopteris dilatata*, *Lonicera periclymenum* and/or *Rubus fruticosus*, and usually *Mnium hornum* in the bryophyte layer. Sphagna, *Deschampsia cespitosa* and *Holcus* spp. are rare or absent.

        W4a *Dryopteris dilatata* – *Rubus fruticosus* sub-community

---

Dryopteris dilatata, *Lonicera* and *Rubus*, typical of W4a, can also be abundant in W6e (*Alnus glutinosa* – *Urtica dioica* woodland, *Betula pubescens* sub-community). W6e usually has more *Urtica dioica*, and less *Molinia caerulea*, than W4a.

W4c can usually be separated from W2b (*Salix cinerea* – *Betula pubescens* – *Phragmites australis*, *Sphagnum* sub-community) by the abundance of *Molinia* and absence/scarcity of *Phragmites australis*.

Sub-community descriptions

W4a *Dryopteris dilatata* – *Rubus fruticosus* sub-community

This sub-community is typical of longer-established and drier stands than the other two, and the downy birch canopy tends to be taller and denser. Other woody species are a little more diverse than is typical for W4. Rowan is frequent and with *Salix cinerea*, young birch and, occasionally, oak can form a shrub layer which is quite dense in parts. *Molinia caerulea*, the dominant field layer species of W4, is very abundant, but often masked by an underscrub of *Rubus fruticosus* and *Lonicera periclymenum*, frequently with *Dryopteris dilatata*. Where there has been local nutrient enrichment or disturbance, this is often marked by the presence of *Chamerion angustifolium*.

W4b *Juncus effusus* sub-community

This is more typical of acidic soligenous mire conditions, and emerging base-poor flushes. Downy birch generally forms a low, open canopy, with alder more frequent than generally in the community. *Salix cinerea* is frequent, but not usually forming a distinct shrub layer, and other woody species are rare. The field layer is generally grassy, with tussocks of sedges and rushes. As well as *Juncus effusus*, *Molinia caerulea* is frequently accompanied by *Deschampsia cespitosa*, *Holcus lanatus* and *Holcus mollis*. Other preferential species include *Carex laevigata*, *Cirsium palustre*, *Hydrocotyle vulgaris*, *Lotus pedunculatus*, *Potentilla erecta* and *Viola palustris*.

W4c *Sphagnum* sub-community

Tends to occur on wetter sites, on deeper peat, than the other two sub-communities. Woody species other than downy birch are rare, with only occasional alder and *Salix cinerea*. The field layer is generally species poor, although quite variable. It often retains an element of the wet heath or mire from which it has developed, and *Calluna vulgaris*, *Erica tetralix*, *Eriophorum angustifolium* and *E. vaginatum* can be frequent. The other distinctive feature is the prominence of Sphagna, forming extensive patches in the wet runnels between *Molinia caerulea* tussocks. *S. fallax* and *S. palustre* are particularly frequent, but several other species can often be found.

W5 *Alnus glutinosa – Carex paniculata* woodland

A community of base-rich, moderately eutrophic, wet to waterlogged organic soils on topogenous or soligenous mires. It is associated with fen peats in open water transitions, flood-plain mires and basin mires where there is strong influence from base-rich ground waters.

Alder and *Salix cinerea* are the most frequent invaders of the preceding swamp and fen communities, and they often co-dominate in young stands, forming low, open and uneven canopies. As the stand ages alder tends to exclude *S. cinerea* or relegates it to the shrub layer. In well-established stands there may be a clearly defined canopy dominated by multi-stemmed alder, over a distinct shrub layer of varying density. As the water level rises, alder may die, so standing dead trees are a common feature. Downy birch may be present in patches with ash and pedunculate oak in drier stands. Hawthorn, holly, rowan, buckthorn, guelder rose and alder buckthorn occur occasionally and the last can be dense in young stands.

The field layer is related to the preceding swamp and fen communities, with a small woodland influence. *Carex paniculata* or *C. acutiformis* usually form a major component, sometimes with *C. elata*, *C. appropinquata*, *C. riparia* and *C. pseudocyperus*. Other prominent fen elements include *Angelica sylvestris*, *Cirsium palustre*, *Eupatorium cannabinum*, *Filipendula ulmaria*, *Iris pseudacorus*, *Lysimachia vulgaris*, *Lythrum salicaria*, *Peucedanum palustre*, *Phragmites australis*, *Urtica dioica*, *Valeriana officinalis* and *V. dioica*. Woodland species, including *Circaea lutetiana* and *Geranium robertianum*, may also occur; and, less often, *Caltha palustris*, *Hydrocotyle vulgaris*, *Mentha aquatica*, *Poa trivialis*, *Ranunculus repens*, *R. flammula* and *Viola palustris*. Sprawling and scrambling species such as *Galium palustre*, *Lonicera periclymenum*, *Ribes nigrum*, *R. rubrum*, *Rosa* spp., *Rubus* spp. and *Solanum dulcamara* may be abundant. Ferns are often conspicuous, especially *Dryopteris dilatata* and, sometimes, *Athyrium filix-femina*, *Dryopteris carthusiana*, *D. cristata*, *Osmunda regalis* and *Thelypteris palustris*. Mosses, including *Brachythecium rutabulum*, *Eurhynchium praelongum*, *Plagiomnium undulatum* and *Rhizomnium punctatum*, are common around sedge tussocks. Sphagna are rare but may occur along base-poor seepage lines where *Pellia epiphylla* is particularly characteristic.

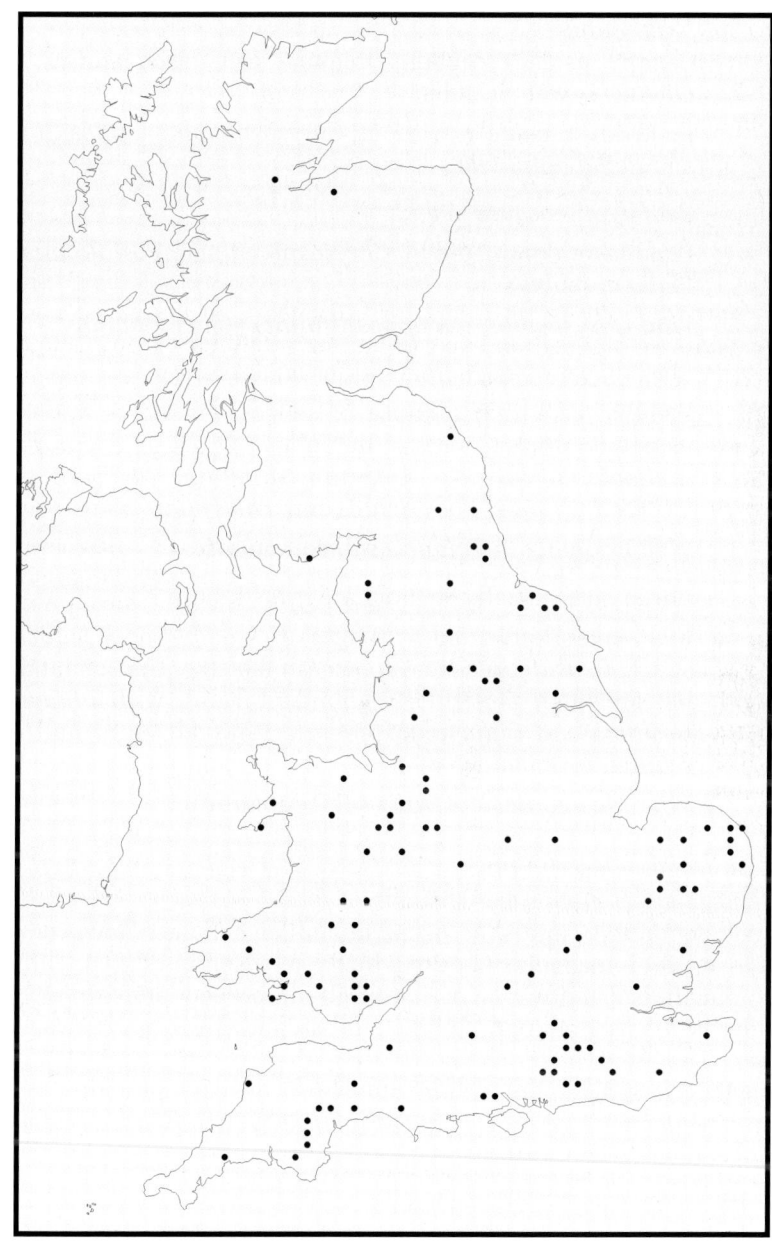

Figure 5 Distribution of W5 *Alnus glutinosa – Carex paniculata* woodland

Key to sub-communities

1 Canopy usually contains frequent ash, birch and/or *Salix cinerea* as well as alder.

Field layer usually without *Chrysosplenium oppositifolium* and *Pellia epiphylla*. ➜ 2

OR Canopy often pure alder.

Field layer dominated by *Chrysosplenium oppositifolium* and *Pellia epiphylla*.

W5c *Chrysosplenium oppositifolium* sub-community

2 (1) Shrub layer often includes alder buckthorn.

Field layer a rich mixture of sedges, tall herbs and ferns, usually including some of the following: *Lycopus europaeus*, *Lysimachia vulgaris*, *Lythrum salicaria*, *Myosotis laxa*, *Osmunda regalis*, *Poa trivialis* and *Thelypteris palustris*.

W5b *Lysimachia vulgaris* sub-community

OR Shrub layer usually without alder buckthorn.

Field layer dominated by sedges, especially *Carex paniculata* and/or *C. acutiformis* (these occur in the other sub-communities, but are often particularly abundant here), with *Phragmites australis* and *Filipendula ulmaria*. The tall herbs and ferns referred to above are rare.

W5a *Phragmites australis* sub-community

Sub-community descriptions
W5a *Phragmites australis* sub-community
This sub-community and W5b occur in similar habitats. However, W5a tends to occur on fragments of mire which have been interfered with, and which remain isolated within agriculturally improved landscapes.

Alder is always dominant, with some ash and silver birch, over *Salix cinerea*. The field layer is poorer than that of W5b and W5c, and dominated by sedges. *Equisetum palustre* is weakly preferential, and *Solanum dulcamara* is more frequent than in W5b and W5c, sometimes forming a dense underscrub.

W5b *Lysimachia vulgaris* sub-community
This is the most species rich of the sub-communities, and can be a prominent feature in topogenous mires which still remain under the close influence of calcareous and eutrophic river waters. Many stands remain

swampy throughout the year, especially where colonisation is recent or where the substrate has started to sink under the weight of the larger trees. The canopy layer is similar to that of W5a, but with a more varied shrub layer, in which alder buckthorn and guelder rose are preferential. In the field layer *Carex remota* and *C. elongata* are preferential. *Cardamine pratensis, Impatiens capensis, Myosotis laxa, Ranunculus flammula, R. repens, Scutellaria galericulata, Thalictrum flavum* and *Viola palustris* are more common than in W5a and W5c. Ferns can be abundant.

W5c *Chrysosplenium oppositifolium* sub-community
This community is typical of springs and seepage lines associated with the emergence of less base-rich water, so calcicoles are not as frequent as in the other two sub-communities. The soils are generally kept very wet, but not surface flooded. They often have a mineral base, with some peat accumulation due to the waterlogged conditions. The tree and shrub layers are simpler and less diverse than in W5a and W5b, dominated by alder with few other species. Hawthorn is absent. *Carex paniculata* is a dominant species with less *C. acutiformis* than in the other two sub-communities. Other sedges and *Phragmites australis* are rare, and tall herbs are less rich, although *Oenanthe crocata* is preferential. Small herbs and bryophytes form a distinct patchy carpet between sedge tussocks, especially *Chrysosplenium oppositifolium* and *Pellia epiphylla*, with some of *Ajuga reptans, Caltha palustris,* and *Cardamine pratensis*. The abundance of *Chrysosplenium* makes this close to W7a. The latter, however, rarely contains the tall sedges typical of W5c.

W6 *Alnus glutinosa – Urtica dioica* woodland

A rather ill-defined community of eutrophic moist soils, especially where there has been substantial deposition of mineral matter, or on flood plain mires where enriched waters flood fen peat.

Alder is usually the most common tree, particularly on wetter soils, but is replaced by *Salix fragilis* in W6b and by downy birch on drier sites. Black poplar may occur, but rarely, and sycamore, ash and pedunculate oak are often occasional species. The shrub layer is usually open and patchy, with *Salix cinerea* the most common shrub, and hawthorn and elder on drier ground. *Salix viminalis*, *S. triandra* and *S. purpurea* are abundant in some stands.

Unlike the other alder types (W5 and W7) the field layer generally lacks tall swamp and fen species. The most typical species is *Urtica dioica*, which can form virtual monocultures, although it may be less common or absent. The few other typical species show a rough transition from wetter to drier habitats. Where soils are moist towards the surface, *Poa trivialis* and *Galium aparine* are frequent, with *Solanum dulcamara* and, often, clumps of swamp and fen species. On drier substrates *Lonicera periclymenum*, *Dryopteris dilatata* and *Rubus fruticosus* are more frequent. Less common species include *Angelica sylvestris*, *Cardamine flexuosa*, *Cirsium palustre*, *Glechoma hederacea* and *Ranunculus repens*. The field layer can appear 'run-down', and may be choked with brushwood from winter flooding or, in the case of drier stands, showing other signs of disturbance.

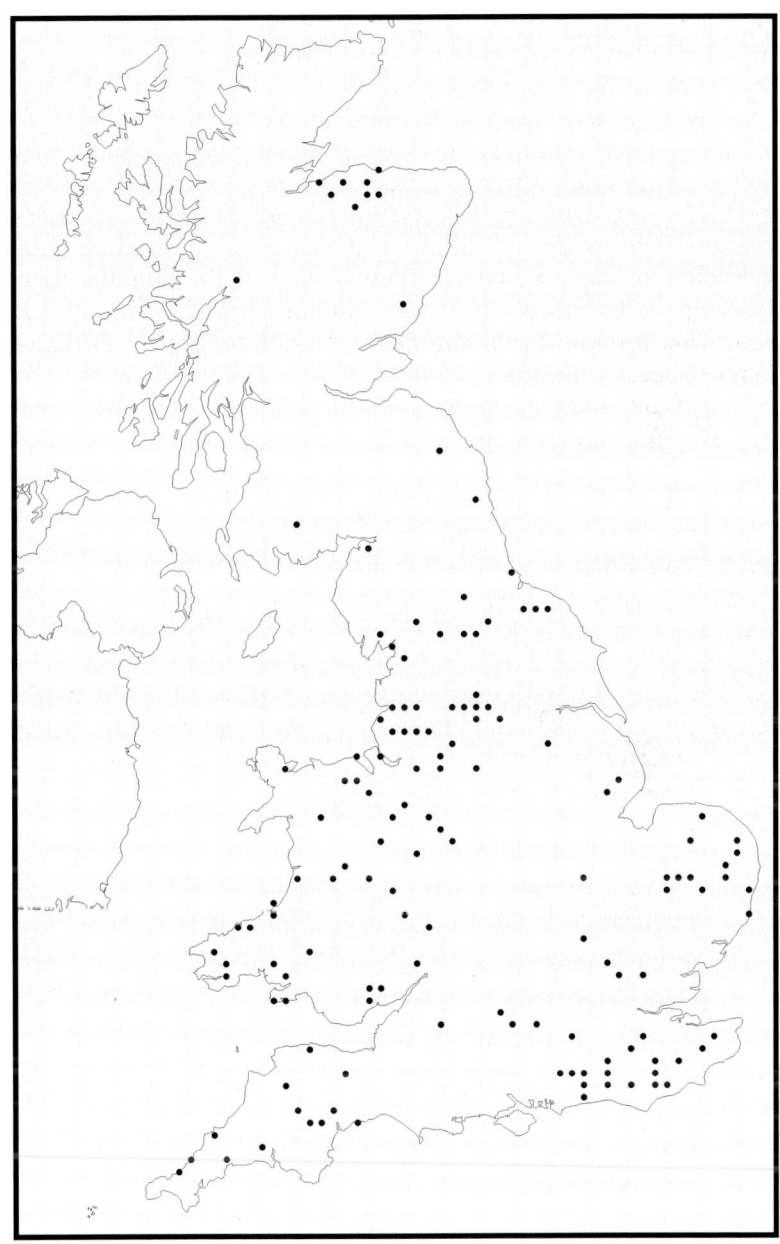

Figure 6 Distribution of W6 *Alnus glutinosa – Urtica dioica* woodland

Key to sub-communities

1	Field layer usually contains *Galium aparine*, *Poa trivialis* or *Solanum dulcamara*, and some of *Carex acutiformis*, *Epilobium hirsutum*, *Equisetum palustre*, *Filipendula ulmaria*, *Galium palustre*, *Iris pseudacorus*, *Phalaris arundinacea* or *Phragmites australis*.

Habitat: Soils often moist towards the surface. ➜ 2

OR Field layer usually contains *Dryopteris dilatata*, *Lonicera periclymenum* or *Rubus fruticosus*, and some of *Circaea lutetiana*, *Dryopteris filix-mas*, *Chamerion angustifolium*, *Geum urbanum*, *Hedera helix*, *Holcus lanatus*, *Rumex obtusifolius* or *Silene dioica*.

Habitat: Drier soils. ➜ 3

2 (1) Osier beds dominated by *Salix triandra*, *S. viminalis* or *S. purpurea* with occasional ash or alder.
W6c *Salix viminalis/S. triandra* sub-community

OR Canopy usually dominated by *Salix fragilis*.

Shrub layer often has *Salix cinerea* and elder.

Field layer with some of *Epilobium hirsutum*, *Galium aparine*, *Iris pseudacorus* and *Phalaris arundinacea*. *Solanum dulcamara* is a common sprawler.
W6b *Salix fragilis* sub-community

OR Canopy dominated by alder.

Shrub layer often with *Salix cinerea* and little else.

Field layer with abundant *Galium aparine*, *Poa trivialis*, *Solanum dulcamara* and/or *Urtica dioica*.
W6a Typical sub-community

3 (1) Shrub layer often with *Salix cinerea* and elder.

Field layer usually with some of *Allium ursinum*, *Circaea lutetiana*, *Dryopteris filix-mas*, *Geum urbanum*, *Hedera helix*, *Rumex obtusifolius* or *Silene dioica*.
W6d *Sambucus nigra* sub-community

OR Canopy with downy birch (or sometimes Scots pine) abundant. Often more so than alder.

Shrub layer with *Salix cinerea* rare or absent. Field layer quite species-poor, sometimes with *Chamerion augustifolium* and *Holcus lanatus*, but with the above species uncommon.

W6e *Betula pubescens* sub-community

Sub-community descriptions

W6a Typical sub-community

This sub-community and the next are characteristic of situations where naturally eutrophic mineral soils are developing by the deposition of rich particulate matter in the slacker reaches of rivers and on flood-plains. Alder dominates the canopy. *Salix fragilis* and osiers are rare, and ash, sycamore and oak occasional. The shrub layer is thin and often composed only of *Salix cinerea*. *Urtica dioica* is usually very abundant, with *Galium aparine*, and often mats of *Poa trivialis*, *Ranunculus repens* or *Glechoma hederacea*. *Arrhenatherum elatius* and *Heracleum sphondylium* are found in drier areas and *Solanum dulcamara* is common. This is similar to W2a (*Salix cinerea – Betula pubescens – Phragmites australis* woodland, *Alnus glutinosa – Filipendula ulmaria* sub-community) but with fewer tall herbs and sedges and more *Poa trivialis* and *Galium aparine*.

W6b *Salix fragilis* sub-community

Alder is frequent, but often as scattered trees in a *Salix fragilis* dominated canopy. *Salix cinerea* and elder are the most frequent shrub layer species, *Salix* being more common in damp situations and elder in drier ones. The shrub layer is often dense with decaying, fallen *Salix fragilis* branches. The field layer contains luxuriant *Urtica dioica* and *Galium aparine*. *Epilobium hirsutum*, *Galium aparine*, *Iris pseudacorus* and *Phalaris arundinacea* may occur and *Solanum dulcamara* can, again, be common, and there may be extensive stretches of sloppy mud.

W6c *Salix viminalis/Salix triandra* sub-community

This can be found in natural situations, where osiers have colonised river islands or fresh alluvium, deposited along the slacker margins of moving waters. Other stands occur in managed or derelict osier beds. The most common species are *Salix triandra*, *S. viminalis* and *S. purpurea*, with occasional emergent alder and ash. The field layer is similar to that in W6a and W6b, with abundant *Urtica dioica* and *Galium aparine* and a ground carpet of *Poa trivialis*.

W6d *Sambucus nigra* sub-community
W6d and W6e are drier than the other sub-communities. They are found on brown alluvial soils and alluvial gleys in situations well removed from the influence of flood waters. W6d is characteristic of more eutrophic and slightly more base-rich situations than W6e. Alder is dominant, with occasional downy birch, and elder and *Salix cinerea* frequent in the shrub layer. *Urtica dioica* and *Galium aparine* are not as prominent as in the wetter sub-communities, and tend to be here replaced by *Dryopteris filix-mas* and *Hedera helix*. *Circaea lutetiana*, *Geum urbanum* and *Mercurialis perennis* are typical where there is local base-enrichment, and *Allium ursinum* and *Petasites hybridus* can dominate locally.

W6e *Betula pubescens* sub-community
Habitat similar to W6d, but less eutrophic and base-rich. This sub-community can develop in semi-ornamental plantings on heavier gleyed soils, or by invading land disturbed by construction work or opencast restoration. Downy birch is often more frequent than alder, and self-seeding or planted pine may be present. The shrub layer is sparse, without *Salix cinerea*. The typical field layer species of the other sub-communities are infrequent, and the most obvious feature is the *Rubus fruticosus/Lonicera periclymenum/Dryopteris dilatata* underscrub. W6e is similar to W4a (*Betula pubescens – Molinia caerulea* typical sub-community), but can be distinguished by the greater abundance of *Urtica dioica* and the lack of Sphagna and *Molinia caerulea*.

W7 *Alnus glutinosa – Fraxinus excelsior – Lysimachia nemorum* woodland

A woodland type typical of moist to very wet mineral soils which are only moderately base-rich and not very eutrophic. It is most extensive in the wetter parts of Britain, but usually occurs in soils where there is no great tendency for peat accumulation.

The canopy is often somewhat open and irregular. Alder is the main woody species, and can be overwhelmingly dominant, but ash is usually common as well. Other species may be frequent, including sycamore and sessile oak where the soil is not permanently moist.

Hazel and hawthorn are the commonest shrubs in drier areas, with *Salix cinerea* more frequent on damper sites. Ash, alder and birch saplings can be common, and bird cherry, blackthorn, elder, holly, rowan and guelder rose can occur.

The field layer is similarly varied, but *Athyrium filix-femina*, *Chrysosplenium oppositifolium*, *Filipendula ulmaria*, *Holcus mollis*, *Lysimachia nemorum*, *Poa trivialis* and *Ranunculus repens* are often present. The wetness and nutrient status of the soil determine what other species may occur. The field layer is usually a low growing cover of herbaceous dicotyledons and grasses: *Lysimachia nemorum*, *Ranunculus repens*, *Poa trivialis* and *Holcus mollis* are most common. *Filipendula ulmaria* and *Athyrium filix-femina* are more scattered but give a layered structure to the herbaceous vegetation. *Juncus effusus* is very frequent in some sub-communities. *Carex remota*, *C. pendula* and *C. laevigata* can occur in some quantity but *C. paniculata* and *C. acutiformis* are rare, giving a good separation between this and other alder – *Carex* types. The local influence of more base-rich waters allows sporadic appearance of *Mercurialis perennis*, *Geum urbanum* and *Circaea lutetiana*. *Rubus fruticosus* frequently forms an underscrub. The bryophyte layer is patchy but *Eurhynchium praelongum* and *Plagiomnium undulatum* are frequent, with some *Thuidium tamariscinum*, *Rhizomnium punctatum* and *Brachythecium rutabulum* and with *B. rivulare* and *Chiloscyphus pallescens* in wetter areas.

Differences between sub-communities are related to the extent of waterlogging, the nature of the water supply and its movement. Some flushes may be a mosaic of types with, for example, W7b in the centre grading into W7c on drier ground. W7c may grade into W9 (*Fraxinus excelsior* – *Sorbus aucuparia* – *Mercurialis perennis* woodland) or W11 (*Quercus petraea* – *Betula pubescens* – *Oxalis acetosella* woodland) around flushes in upland woods. The abundance of alder in the canopy is generally a good indicator of W7, but many stands are small and alder may be sparse or even absent on occasions. W7 is less exclusively western than indicated by the originally published distribution map (Rodwell 1991).

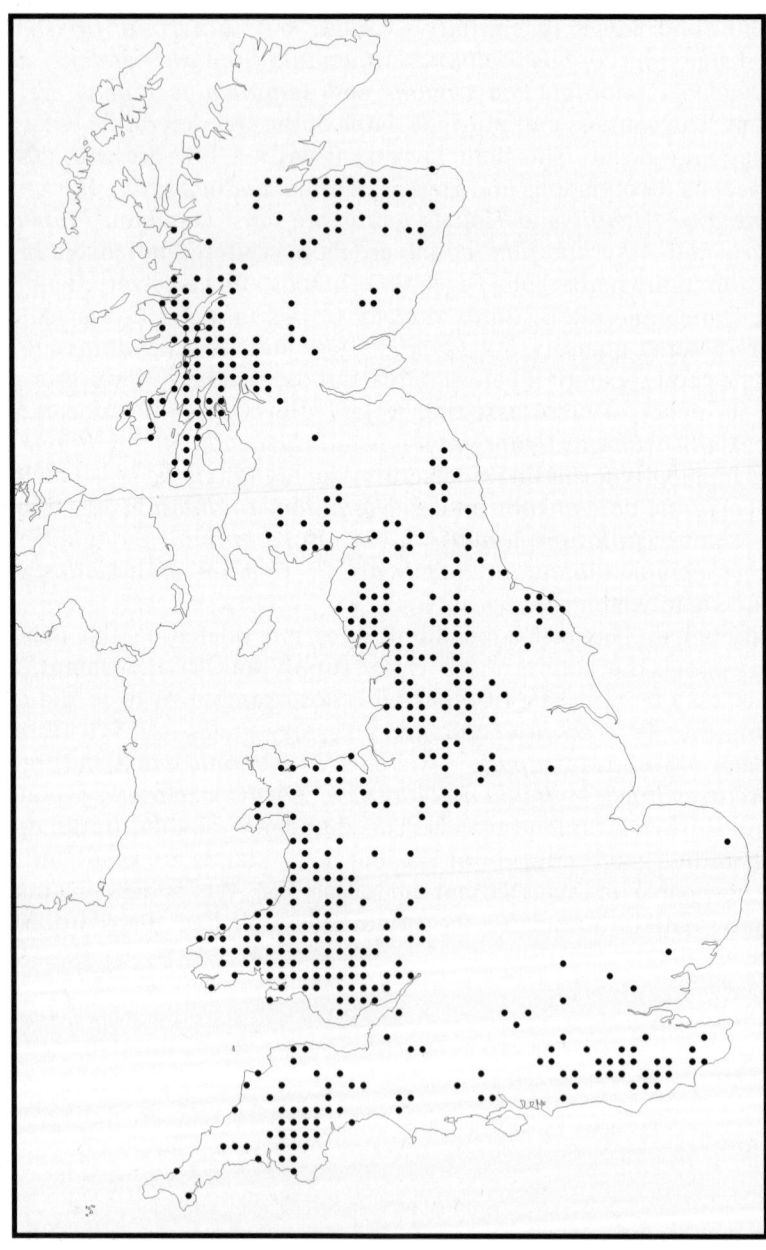

Figure 7 Distribution of W7 *Alnus glutinosa – Fraxinus excelsior – Lysimachia nemorum* woodland

Key to sub-communities

1   Shrub layer patchy, hazel and hawthorn uncommon.

           Field layer with *Ranunculus repens* and *Chrysosplenium oppositifolium* commonly forming a carpet. Amongst the taller herbs, *Angelica sylvestris*, *Galium aparine* and/or *Urtica dioica* are dominant with, less frequently, *Phalaris arundinacea*.

           W7a *Urtica dioica* sub-community

OR   Shrub layer often contains hazel and hawthorn.

           Field layer without *Chrysosplenium*, or if with, then without the above mixtures of species.   ➔ 2

2 (1)   Field layer variable and often rich. Dominants can include *Carex laevigata*, *C. pendula*, *C. remota*, *Equisetum telmateia* or *Juncus effusus*. *Cirsium palustre* and *Valeriana officinalis* can be frequent and bryophytes, including *Brachythecium rivulare*, *Calliergonella cuspidata*, *Chiloscyphus pallescens* and *Palustriella commutata*, are often abundant.

           W7b *Carex remota – Cirsium palustre* sub-community

OR   Field layer usually with abundant *Deschampsia cespitosa* and *Dryopteris dilatata*, and often *Oxalis acetosella*. Bryophytes, especially *Atrichum undulatum* and *Mnium hornum*, are often common.

           W7c *Deschampsia cespitosa* sub-community

Sub-community descriptions

W7a *Urtica dioica* sub-community
This sub-community occurs on light-textured alluvial soils, on flat or gently-sloping terraces of young river systems. The soils are free-draining, but kept moist by high water tables or by flushing from above. Alder is generally dominant but ash and, less frequently, sycamore can be quite common. Birch is less frequent than is typical for the rest of the community. The understorey is sparse; elder and *Salix cinerea* are generally the only common shrubs, although saplings, principally of sycamore and ash, can be frequent. The field layer typically consists of a carpet of *Chrysosplenium oppositifolium* and *Ranunculus repens*, with *Urtica dioica* and often *Angelica sylvestris*, *Filipendula ulmaria*, *Galium aparine* and/or *Lysimachia nemorum*.

W7b *Carex remota – Cirsium palustre* sub-community
This is associated with springs or seepage lines, where groundwater

Key to sub-communities

1 — Canopy may have frequent lime, hornbeam or pedunculate oak, but sycamore and wych elm are rare.

Shrub layer may have abundant Midland hawthorn.

Field layer with some of *Anemone nemorosa, Deschampsia cespitosa, Glechoma hederacea, Hedera helix, Poa trivialis, Ranunculus ficaria*.

Habitat: typical of heavy base-rich mull soils, mainly in the south-east. → 2

OR — Canopy often with frequent sycamore, wych elm or sessile oak. Pedunculate oak rarer.

Field layer with some of *Allium ursinum, Geranium robertianum, Teucrium scorodonia* frequent or abundant, and the species listed above rare.

Habitat: light, well-drained but moist soils in the north-west. → 3

This division based on the tree and shrub layer composition should be used only as an indication. If necessary, all seven sub-communities should be considered.

2 (1) — Field layer with abundant *Hedera helix* and often *Brachypodium sylvaticum*, but with few other species (cf. W8e). *Hedera* frequently forming a dense ground carpet.

W8d *Hedera helix* sub-community

OR — Field layer with abundant, often dominant *Deschampsia cespitosa* and some *Filipendula ulmaria, Juncus effusus, Lysimachia nemorum* or *Potentilla sterilis*. *Mercurialis perennis* is often sparse compared to the other sub-communities.

W8c *Deschampsia cespitosa* sub-community

OR — Field layer with *Anemone nemorosa* and *Ranunculus ficaria* abundant or dominant in spring, often with *Lamiastrum galeobdolon* or *Rumex sanguineus*. Late in the season this may be difficult to distinguish from W8a.

W8b *Anemone nemorosa* sub-community

OR — Field layer with some of *Ajuga reptans, Glechoma hederacea, Poa trivialis, Primula* spp. or *Viola riviniana/reichenbachiana*.

W8a *Primula vulgaris* – *Glechoma hederacea* sub-community

3 (1)		Field layer dominated by *Allium ursinum* in spring. Later in the year it may be largely bare soil.

W8f *Allium ursinum* sub-community

OR Field layer with abundant *Hedera helix* and some of *Galium aparine*, *Geranium robertianum*, *Phyllitis scolopendrium*, *Urtica dioica*. Bryophytes often abundant including *Ctenidium molluscum*, *Eurhynchium striatum*, *Thamnobryum alopecurum*. *Allium ursinum* rare.

W8e *Geranium robertianum* sub-community

OR Shrub layer often rich, with some of dogwood, buckthorn, rowan, guelder rose as well as field maple, hazel or hawthorn.

Field layer often a complex mosaic with five or more of *Angelica sylvestris*, *Arrhenatherum elatius*, *Campanula latifolia*, *C. trachelium*, *Deschampsia cespitosa*, *Melica nutans*, *Melica uniflora*, *Polystichum aculeatum*, *Sanicula europaea*, *Teucrium scorodonia* or *Viola riviniana*.

W8g *Teucrium scorodonia* sub-community

Sub-community descriptions

W8a, b and c have a generally south-eastern distribution. They are particularly common in woods which have been managed as coppice-with-standards. The canopy/shrub layer structure of high forest is often absent, although this is changing where coppicing has been abandoned. Pedunculate oak is the most common woody species after ash, maple and hazel, and is strongly preferential to this group. Hazel is the most frequent shrub, although hawthorns are common, and Midland hawthorn is preferential, particularly in long-established stands. Other species may dominate locally, including small-leaved lime, hornbeam and invasive elms. Lime and hornbeam can form dense, single species stands, often accentuated by generations of coppicing. Field layer species include *Anemone nemorosa*, *Deschampsia cespitosa*, *Glechoma hederacea*, *Poa trivialis* and *Primula vulgaris* (*P. elatior* in East Anglia). The abundance of *Mercurialis perennis*, and the type of sub-community, varies according to the duration and extent of soil waterlogging.

W8a *Primula vulgaris – Glechoma hederacea* sub-community
This is the most common sub-community. Lime and hornbeam can be locally abundant. The ground flora is dominated by *Mercurialis perennis* (except where grazed out), with frequent *Ajuga reptans*, *Glechoma*

hederacea, Poa trivialis, Primula spp. *Hyacinthoides non-scripta* is more prominent on damper soils.

W8b *Anemone nemorosa* sub-community
This sub-community becomes more common where soils remain wetter for longer in spring, on heavy clay soils in the south-east, and locally on wet sites in the north-west. *Anemone nemorosa* and *Ranunculus ficaria* are vernal dominants and *Hyacinthoides non-scripta* is often more abundant than *Mercurialis perennis*. The few other preferentials include *Carex acutiformis, C. pendula, C. remota, C. strigosa* and *Rumex sanguineus*. Separation of W8a and W8b can be difficult in late summer where *Anemone nemorosa* and *Ranunculus ficaria* have died back, and stands may have to be left as W8a/b.

W8c *Deschampsia cespitosa* sub-community
Characteristic of heavy, wet, often trampled soils, which are free from water-logging for only a short period in the summer. Abundant *Deschampsia cespitosa* is the most obvious feature, especially in open conditions, such as young coppice and rides. *Mercurialis perennis* and *Hyacinthoides non-scripta* are less common than in the other sub-communities. In disturbed situations diversity is often increased by ruderal species, especially *Cirsium* spp., *Chamerion angustifolium, Epilobium* spp., *Hypericum* spp., *Juncus conglomeratus, J. effusus* and *Rumex* spp.

W8e, f and g are more common to the north and west. Sessile oak and oak hybrids are more abundant here, as are wych elm and sycamore. A high forest structure is more common than in W8a-c. Waterlogging plays a lesser role in the distinctions between these sub-communities. Species of clay soils (e.g. *Hyacinthoides non-scripta*) give way to those more typical of free-draining soils, like *Brachypodium sylvaticum* and *Geranium robertianum*.

W8e *Geranium robertianum* sub-community
The central type of the group. A eutrophic type, characterised by species such as *Urtica dioica, Galium aparine* and *Geranium robertianum*. Ash is the most common canopy species, with some birch, cherry, yew and whitebeam. Hazel, elder, hawthorn and holly are the main shrubs and there is often a *Rubus fruticosus* underscrub. *Mercurialis perennis* is abundant with *Brachypodium sylvaticum* and *Hedera helix*. Ferns such as *Polystichum setiferum* and *Phyllitis scolopendrium* are common, but not *Athyrium filix-femina, Dryopteris filix-mas* or *D. dilatata* which are indicative of W9, and there is often high bryophyte cover, frequently including *Brachythecium rutabulum, Eurhynchium praelongum, E. striatum* or *Thuidium tamariscinum*.

W8f *Allium ursinum* sub-community
This is found on deeper, moister slope-foot soils. The prominence of *Allium ursinum*, which often forms a continuous carpet, is the most distinctive feature. *Mercurialis perennis* may be more prominent later in the year.

W8g *Teucrium scorodonia* sub-community
More species-rich, with greater structural complexity, than the other types. Sycamore and wych elm are less prominent but oak, yew, beech and limes may be present. The shrub layer is prominent and varied, often with dogwood, guelder rose, rowan and buckthorn. The field layer reflects a patchy woody cover, a complex rocky topography and moderately montane climatic conditions. It often includes *Deschampsia cespitosa*, *Teucrium scorodonia*, *Campanula latifolia*, *C. trachelium*, *Melica uniflora*, *Sanicula europaea*, *Viola riviniana*. Less commonly, *Convallaria majalis*, *Geranium sanguineum*, *Melica nutans*, *M. uniflora*, *Myosotis sylvatica* or *Polygonatum* spp. may occur. Bryophyte cover is often low because of the drier conditions.

W8d is largely southern in its distribution, but overlaps the ranges of both the above groups. This is reflected in its tree and shrub layer. In the south-east it is often associated with recent woodland and hedgerows; in the west it is also found in ancient woods.

W8d *Hedera helix* sub-community
The structure is often simple, with a dense canopy and shrub layer of ash or oak over hawthorn, hazel and/or sparse field maple. The ground flora is species-poor and dominated by *Hedera helix* with some *Geum urbanum*, *Circaea lutetiana*, *Brachypodium sylvaticum* and *Poa trivialis*.

W9 *Fraxinus excelsior* – *Sorbus aucuparia* – *Mercurialis perennis* woodland

A community of permanently moist calcareous soils in the sub-montane climate of north-west Britain. It is commonly found by streams and flush lines in the uplands, where the climate is cool, wet and windy, and hence unsuitable for the more continental species found in south-eastern mixed deciduous woods (W8, W10). Winter temperatures are comparatively mild and this, combined with high humidity, helps give the community a markedly oceanic and winter-green character with an abundance of ferns and bryophytes.

Ash and hazel are the most abundant woody species, and downy birch and rowan may be co-dominant. The more continental trees and shrubs characteristic of W8 (small-leaved lime, hornbeam, Midland hawthorn) are usually absent. The community varies from well-developed ash, wych elm, sycamore and sessile oak high forest with a distinct shrub layer, to scrubby mixtures of hazel, downy birch and rowan with scattered ash trees in the far north-west, and in exposed areas with irregular topography.

The field layer is usually a complex mosaic, with no single species dominating. The pattern may be further complicated by the effects of local flushing. *Mercurialis perennis* and *Hyacinthoides non-scripta* are both frequent and *Brachypodium sylvaticum*, *Circaea lutetiana*, *Geranium robertianum* and *Geum urbanum* may all be common. *Primula vulgaris*, *Poa trivialis* and *Deschampsia cespitosa*, uncommon in north-western forms of W8, are more common in W9, whereas *Urtica dioica* and *Galium aparine* occur only occasionally. Ferns, such as *Athyrium filix-femina*, *Blechnum spicant*, *Dryopteris dilatata* and *D. filix-mas* can be prominent. Other features distinctive of W9 include an abundance of *Oxalis acetosella*, a grassy appearance (*Arrhenatherum elatius*, *Brachypodium sylvaticum*, *Deschampsia cespitosa* and *Poa trivialis*), and a well-developed bryophyte layer. *Atrichum undulatum*, *Eurhynchium praelongum*, *E. striatum*, *Mnium hornum*, *Plagiomnium undulatum*, and *Thuidium tamariscinum* are often common, with occasional *Cirriphyllum piliferum*, *Hypnum cupressiforme*, *Lophocolea bidentata*, *Plagiochila asplenioides* and *Rhytidiadelphus triquetrus*.

In many upland woods small areas of W9 occur at the base of slopes or along flush lines. There may be a graduation from 'oak' communities (W11, W17) through W9 to W7 *Alnus glutinosa – Fraxinus excelsior – Lysimachia nemorum* woodland over quite short distances. The absence of shrubs with a southern distribution (e.g. dogwood and spindle) and the greater frequency of rowan helps to separate the community from its southern counterpart, W8 (*Fraxinus excelsior – Acer campestre – Mercurialis perennis* woodland).

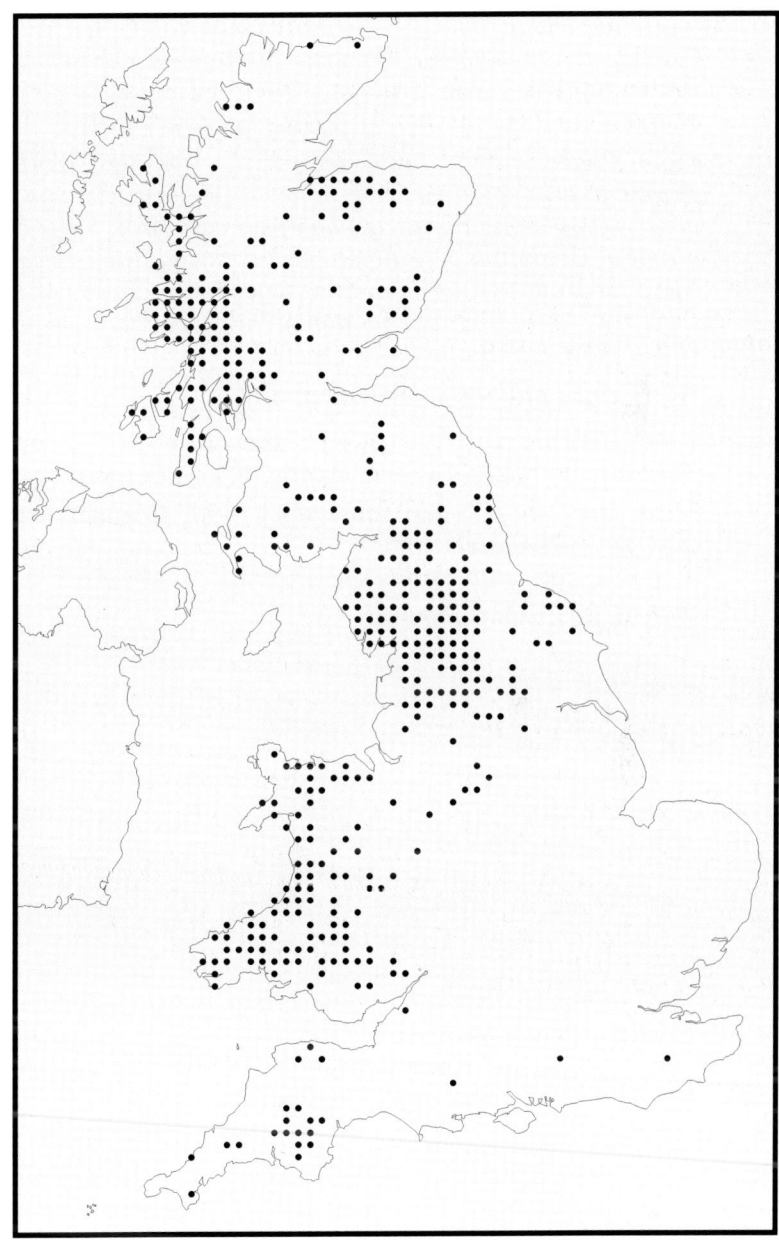

Figure 10 Distribution of W9 *Fraxinus excelsior – Sorbus aucuparia – Mercurialis perennis* woodland

Key to sub-communities

Canopy a rather open, scrubby mixture of downy birch, rowan and hazel with frequent ash and locally abundant bird cherry.

Field layer often luxuriant with five or more of *Arrhenatherum elatius, Cirsium heterophyllum, Conopodium majus, Crepis paludosa, Deschampsia cespitosa, Filipendula ulmaria, Geranium sylvaticum, Geum rivale, Rumex acetosa, Vicia sepium.* Other common species include *Agrostis capillaris, Anthoxanthum odoratum, Brachypodium sylvaticum, Holcus* spp. and *Ranunculus repens.*

W9b *Crepis paludosa* sub-community

OR   Canopy with a more typical high forest or neglected coppice structure, dominated by ash, sycamore, and/or wych elm.

Field layer shorter and more open with frequent *Circaea lutetiana, Dryopteris dilatata, Geum urbanum, Mercurialis perennis, Potentilla sterilis.*

W9a Typical sub-community

Sub-community descriptions

W9a Typical sub-community

Ash and hazel are the most abundant woody species with, frequently, a high forest structure. Elm, sycamore and sessile oak are often present, but downy birch and rowan are less frequent. Hawthorn is the most common shrub after hazel, in an often well-defined shrub layer.

The field layer is variable. Grasses are less important than in W9b and ferns often predominate, especially *Dryopteris* spp. and *Athyrium filix-femina. Geum urbanum, Circaea* spp. and *Potentilla sterilis* are preferential, and other locally abundant species include *Galium aparine, Geranium robertianum, Hyacinthoides non-scripta, Mercurialis perennis, Oxalis acetosella, Rubus fruticosus, Urtica dioica, Viola riviniana. Lamiastrum galeobdolon* and *Arum maculatum* can be more common in the south. Bryophytes are often very abundant.

W9b *Crepis paludosa* sub-community

Ash, birch and rowan predominate, over a patchy hazel understorey, with few other woody species. *Mercurialis perennis* is still frequent, but much less abundant than in W9a, and ferns are much less important. Tall herbs and grasses predominate, in a luxuriant field layer, which often includes *Agrostis capillaris, Anthoxanthum odoratum, Arrhenatherum elatius,*

Brachypodium sylvaticum, Cirsium heterophyllum, Conopodium majus, Crepis paludosa, Dactylis glomerata, Deschampsia cespitosa, Holcus spp., *Filipendula ulmaria, Geum rivale* (replacing *G. urbanum* here), *Poa trivialis, Rumex acetosa* and *Vicia sepium*. Bryophytes are again abundant, the larger pleurocarps being distinctive.

W10 *Quercus robur – Pteridium aquilinum – Rubus fruticosus* woodland

A community of base-poor brown earths mainly in the lowlands of southern Britain. It has a slight continental element, which differentiates it from north-western types, but it is more oceanic than similar European types.

Oak is the most common tree (usually pedunculate in England, but hybrids predominate in Wales), and silver birch is abundant, especially in younger stands. Ash tends to be rare in south-eastern stands, but can be more frequent, with sycamore and sometimes wych elm, in the north-west. Small-leaved lime, hornbeam and sweet chestnut are locally prominent. Other species, present at low frequencies, include holly, beech, wild cherry, wild service and crab apple with alder and aspen on damper soils. Some stands are dominated by planted conifers, but there is often enough of a field layer to classify the type. Hazel is the most abundant shrub, often with hawthorn.

The field layer lacks base-rich indicators such as *Mercurialis perennis*. *Hyacinthoides non-scripta* and *Anemone nemorosa* are often spring dominants, but *Rubus fruticosus, Pteridium aquilinum* and/or *Lonicera periclymenum* are the most common species. *Dryopteris filix-mas* and *D. dilatata* may be locally abundant and conspicuous where *Pteridium aquilinum* is sparse. Grasses, including *Deschampsia cespitosa, Holcus mollis, Melica uniflora, Milium effusum* or *Poa trivialis*, can be common, especially before the emergence of *Pteridium* fronds, although this is more a feature of W11. A wide range of other species occur locally including *Ceratocapnos claviculata, Digitalis purpurea, Luzula pilosa, Silene dioica, Solidago virgaurea* and *Stellaria holostea*. Bryophyte cover is generally low, although *Eurhynchium praelongum* and *Mnium hornum* can be abundant.

Bracken-dominated stands may be difficult to separate from W16 (*Quercus* spp. – *Betula* spp. – *Deschampsia flexuosa* woodland), but this tends to be more 'heathy' with *Calluna vulgaris, Deschampsia flexuosa* or *Vaccinium myrtillus*.

Stands with some beech may be difficult to separate from W14 (*Fagus sylvatica – Rubus fruticosus* woodland). A subjective decision may be needed depending on the proportion of beech and its origin. Oak stands recently underplanted with beech will usually be described as W10.

Ash and sometimes sycamore may be part of the canopy, particularly in W10c and W10e. These may be separated from W8/9 stands by the scarcity of *Mercurialis perennis* and other basophilous herbs and grasses. W10e, with locally abundant *Holcus mollis*, may be difficult to separate from W11, but the latter usually contains more *Agrostis* spp., *Anthoxanthum odoratum* and particularly *Deschampsia flexuosa*.

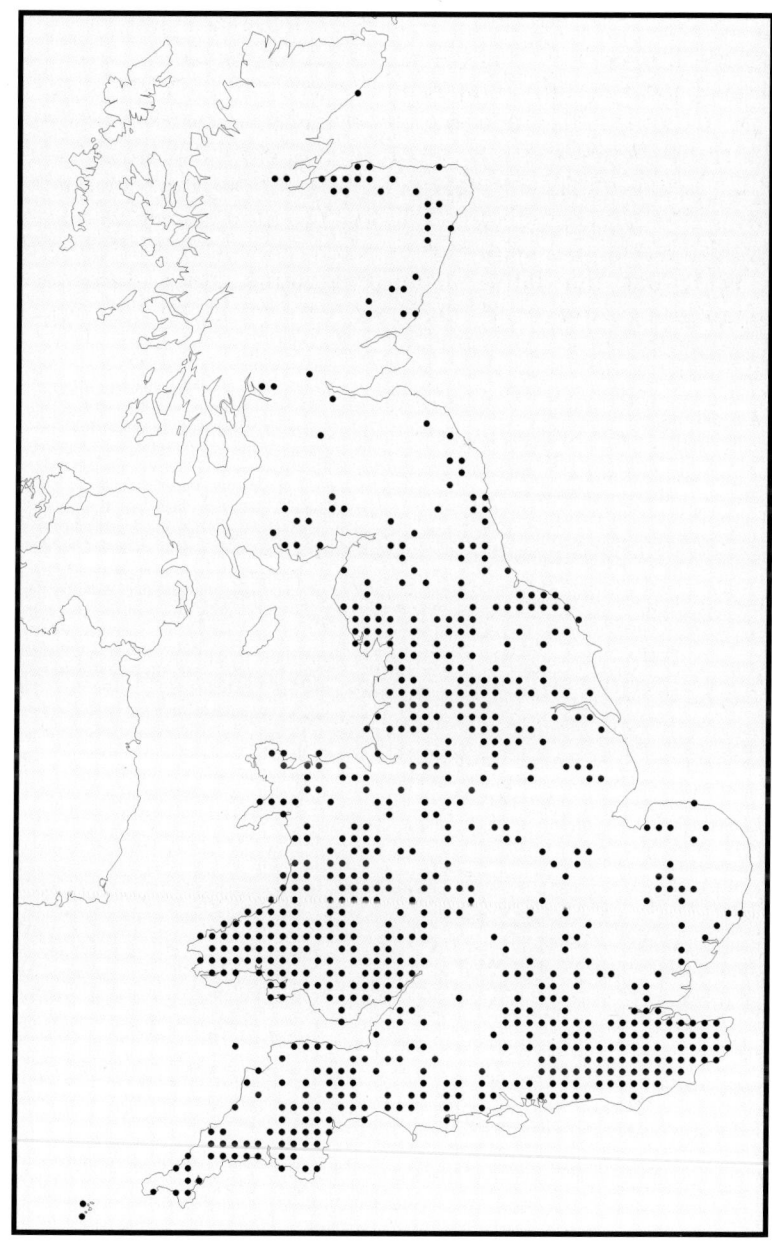

Figure 11 Distribution of W10 *Quercus robur – Pteridium aquilinum – Rubus fruticosus* woodland

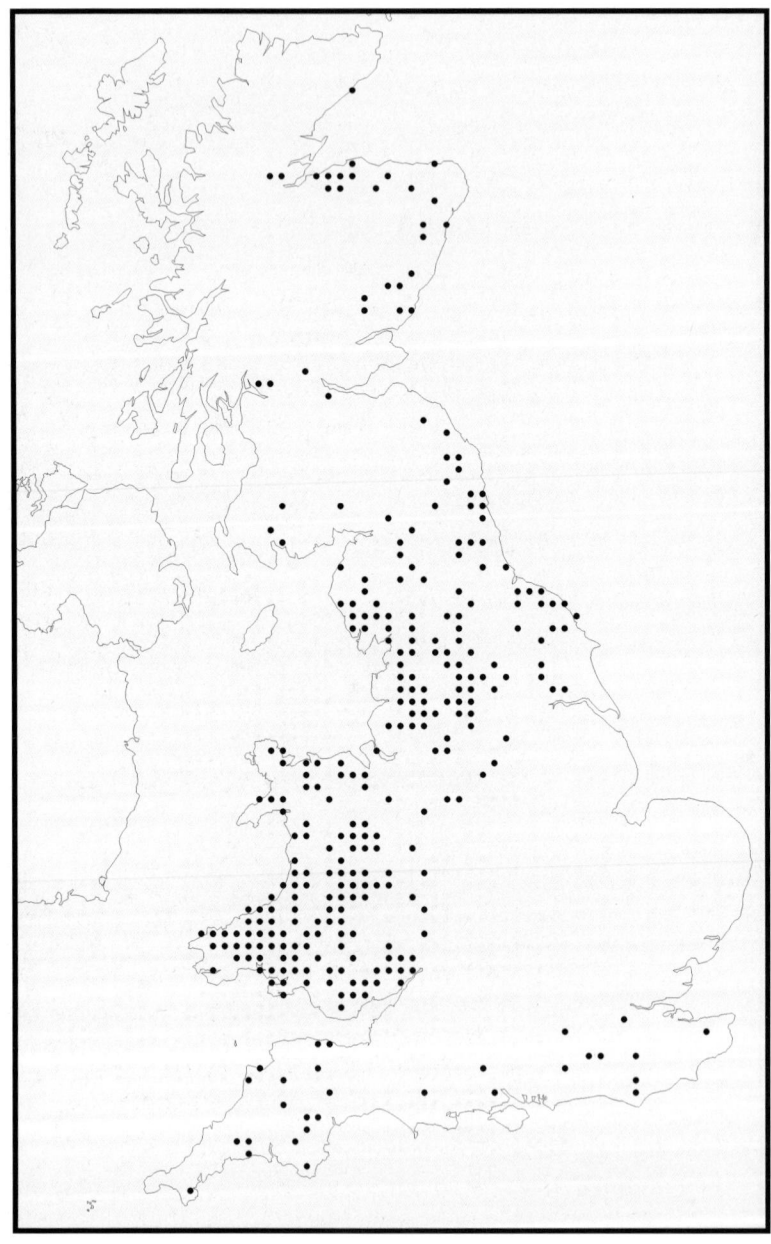

Figure 12 Distribution of W10e *Acer pseudoplatanus – Oxalis acetosella* sub-community woodland

Key to sub-communities

	Field layer dominated by *Anemone nemorosa* in spring. **W10b *Anemone nemorosa* sub-community**
OR	Field layer with *Hedera helix* forming a prominent carpet. **W10c *Hedera helix* sub-community**
OR	Canopy often planted with conifers or oak. Field layer dominated by *Holcus lanatus*. **W10d *Holcus lanatus* sub-community**
OR	Canopy often with some sycamore or ash. Sessile (rather than pedunculate) oak may be dominant. Field layer with some of *Dryopteris dilatata*, *Holcus mollis*, *Oxalis acetosella*. Bryophytes more prominent than in the other sub-communities. **W10e *Acer pseudoplatanus – Oxalis acetosella* sub-community**
OR	Field layer with stands of *Hyacinthoides non-scripta*, *Lonicera periclymenum*, *Pteridium aquilinum* and/or *Rubus fruticosus* and few or none of the species that distinguish the above sub-communities. **W10a Typical sub-community**

Sub-community descriptions

W10a Typical sub-community

W10a is the 'central' sub-community, and acts as the 'default' type from which other sub-communities are separated by abundance of particular species or groups of species (cf W8a). Generally dry oak/birch woods, although oak is sometimes excluded by management, as in coppices of hazel, lime, hornbeam and chestnut, mainly in the south. The most common shrub is hazel, with some hawthorn, holly, rowan, guelder rose, apple or elder. *Hyacinthoides non-scripta* is dominant in spring usually with *Rubus fruticosus*, *Pteridium aquilinum* and/or *Lonicera periclymenum* abundant later.

W10b *Anemone nemorosa* sub-community

This sub-community is more common on winter or spring waterlogged soils on the heavier clays, e.g. on waterlogged plateaus and in hollows. Pedunculate oak is usually dominant, with some birch, over a thin hazel understorey. Sweet chestnut can be abundant, and although lime and hornbeam are sparse associates, they can dominate locally. A carpet of

Anemone nemorosa is the most distinctive feature in spring. *Pteridium aquilinum* is less abundant than in the community as a whole, and soils are generally too moist for *Hyacinthoides non-scripta* to dominate.

W10c *Hedera helix* sub-community
A sub-community of the more Atlantic western side of Britain. Oak is the most common tree, with some beech and ash over a sparse understorey of hazel, hawthorn or elder. Birch is rarer than in the community as a whole, and holly is more abundant than in the other sub-communities. The most distinctive feature is the carpet of *Hedera helix*. *Pteridium aquilinum* is frequent but often at low cover. *Rubus fruticosus* and *Lonicera periclymenum* are abundant, but *Hyacinthoides non-scripta* is uncommon. There may be occasional patches of *Dryopteris dilatata*, *D. filix-mas*, *Galium odoratum* or *Holcus* spp. It may sometimes be transitional to W14.

W10d *Holcus lanatus* sub-community
A species-poor sub-community, typical of oak and conifer plantations and of recent secondary birch/oak woods. The understorey is sparse or absent, with infrequent hazel, but scattered hawthorn, elder or blackthorn. *Holcus lanatus* is the most distinctive feature, usually with tall herb/ruderal/ephemeral species, such as *Senecio jacobaea* and *Dactylis glomerata*. *Pteridium aquilinum* is often abundant, and *Rubus fruticosus* and *Lonicera periclymenum* can be common, but there are generally few other species. *Hyacinthoides non-scripta* is very rare, and the other normal associates of W10 are uncommon.

W10e *Acer pseudoplatanus – Oxalis acetosella* sub-community
This is the most oceanic sub-community and hence more common in the north-west. It may be difficult to separate from W11 (*Quercus petraea – Betula pubescens – Oxalis acetosella* woodland) particularly in north-east England. The latter, however, tends to have a much more grassy appearance with *Agrostis capillaris*, *Anthoxanthum odoratum* and *Deschampsia flexuosa*. In Wales and western England, W10a (the typical sub-community) stands often contain a scattering of species preferential to W10e (including *Dryopteris dilatata*, *Eurhynchium praelongum* and *Holcus mollis*), and these two sub-communities can be difficult to separate. W10e will usually have two or more preferentials, at least one constant through the stand.

Sessile, pedunculate and hybrid oaks are present with ash, sycamore and some wych elm, in a (usually) high forest structure. Silver birch is quite sparse, and hornbeam and lime absent or rare. Hazel is the most common shrub, with some hawthorn or holly. The ground flora can be quite rich, and *Oxalis acetosella*, *Viola riviniana* and a good bryophyte cover are the most distinctive features. *Athyrium filix-femina*, *Dryopteris dilatata* and *Holcus mollis* are more common than in the rest of the community. Often grades into W16 on free-draining acidic soils.

W11 *Quercus petraea – Betula pubescens – Oxalis acetosella* woodland

A community of moist, free-draining (but not excessively leached) base-poor brown earth soils in the cooler, wetter north-west of Britain. It is characteristic of substrates that are neither markedly calcareous nor strongly acidic. The character of the community is heavily influenced by grazing.

Oak (usually sessile, although pedunculate and hybrids may occur) and/or downy birch (especially at higher altitudes and in the extreme north-west) usually dominate. Silver birch can be frequent, particularly in the east, and hybrid birches are present throughout the community. Oak-dominated stands often have a tall, well-grown high-forest canopy, or are derived from coppice with a low cover of crooked, multi-stemmed individuals. Birch-dominated stands usually have a more open canopy, often consisting of widely spaced, rather moribund trees. Other tree species are scarce. Rowan and hazel may be locally common, but the shrub layer is generally less well developed than in W10, and regeneration is often limited due to excessive grazing.

Grasses are important in the field layer, particularly *Agrostis vinealis*, *A. capillaris*, *Anthoxanthum odoratum*, *Deschampsia flexuosa* and *Holcus mollis*, although they may be reduced where *Pteridium aquilinum* is dense. *Hyacinthoides non-scripta* is a vernal dominant in western stands but *Anemone nemorosa* is more frequent on moister soils and in the more continental regions of north-east Scotland. *Oxalis acetosella* and *Viola riviniana* are characteristic of such permanently moist soils and occur with species indicative of surface-leached soils, such as *Galium saxatile* and *Potentilla erecta*. Other characteristic herbs include *Conopodium majus*, *Stellaria holostea* and *Teucrium scorodonia*, but by mid-summer many stands are dominated by *Pteridium aquilinum*. *Lonicera periclymenum* and *Rubus fruticosus* may be abundant, particularly in ungrazed stands, as can *Luzula sylvatica*, although such stands may be better referred to W10e. Ferns can be conspicuous, especially *Blechnum spicant* and *Oreopteris limbosperma* with less frequent *Athyrium filix-femina*, and *Dryopteris* spp. Species characteristic of very acid soils, such as *Calluna vulgaris*, *Vaccinium* spp. and *Erica* spp., tend to be scarce.

Bryophytes are often common, particularly in sheltered north-facing slopes and ravines, although not as predominant as in W17, except on or around rocks. Characteristic species include *Dicranum majus, Hylocomium splendens, Lophocolea bidentata, Pleurozium schreberi, Polytrichum formosum, Scleropodium purum, Rhytidiadelphus squarrosus, R. triquetrus* and *Thuidium tamariscinum*.

Bracken-dominated stands may be difficult to separate from W16 (*Quercus* spp. – *Betula pendula* spp. – *Deschampsia flexuosa* woodland) or even W10 (*Quercus robur* – *Pteridium aquilinum* – *Rubus fruticosus* woodland). Some of the other grasses, herbs and bryophytes characteristic of W11 will usually be present, even if at low frequency. Division solely on the basis of the oak/birch species should not be used.

On more calcareous and base-rich substrates to the west and north W11a often grades to W9a (typical sub-community of the *Fraxinus excelsior* – *Sorbus aucuparia* – *Mercurialis perennis* woodland) but in the latter sycamore, ash and wych elm are almost always more abundant than sessile oak. The grasses characteristic of W11 are all rare in W9 whereas *Mercurialis perennis* and other base-rich herbs are prominent.

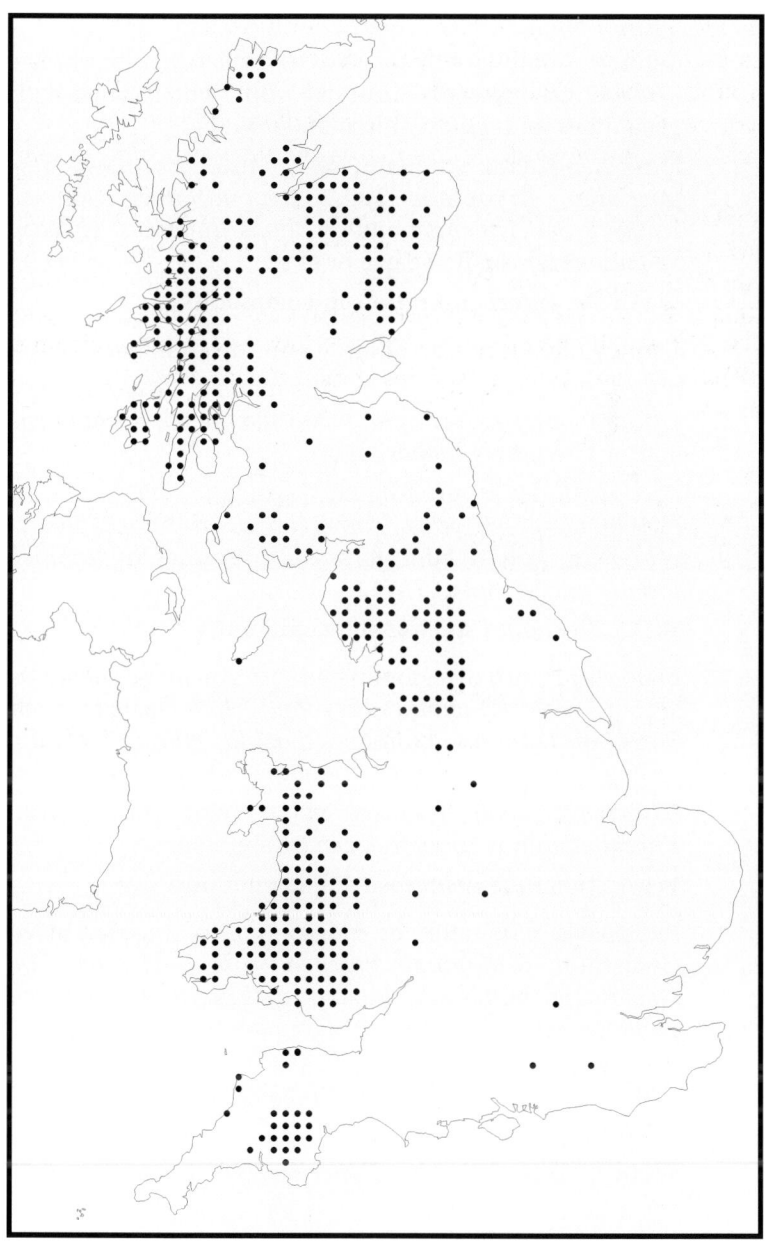

Figure 13 Distribution of W11 *Quercus petraea – Betula pubescens – Oxalis acetosella* woodland

Key to sub-communities

Canopy usually dominated by oak rather than birch, and ash is sometimes present. Shrub layer often dominated by hazel, sometimes with downy birch and rowan.

Field layer with abundant *Deschampsia cespitosa*, *Digitalis purpurea*, *Dryopteris* spp., *Hyacinthoides non-scripta*, *Lonicera periclymenum* and/or *Rubus fruticosus*. *Hylocomium splendens* is usually sparse or absent.

W11a *Dryopteris dilatata* sub-community

OR Canopy often a thicket of birch and rowan rather than oak.

Field layer with *Athyrium filix-femina*, *Blechnum spicant*, *Hyacinthoides non-scripta*, *Potentilla erecta*, *Primula vulgaris* and/or *Oreopteris limbosperma*.

Bryophyte layer prominent, especially *Dicranum majus*, *Hypnum cupressiforme*, *Isothecium myosuroides*, *Pleurozium schreberi*, *Polytrichum formosum* and/or *Rhytidiadelphus loreus*. This is close to W17.

W11b *Blechnum spicant* sub-community

OR Field layer usually dominated by *Anemone nemorosa* in spring, with *Hyacinthoides non-scripta* sparse or absent. Some of *Lathyrus linifolius*, *Luzula pilosa*, *Melampyrum pratense*, *Rubus idaeus* or *Trientalis europaea* also present.

Bryophyte layer less extensive than above, although *Rhytidiadelphus triquetrus* can be abundant.

W11c *Anemone nemorosa* sub-community

OR Field layer with some of *Ajuga reptans*, *Angelica sylvestris*, *Cerastium fontanum*, *Festuca rubra*, *Holcus lanatus*, *Hypericum pulchrum*, *Luzula multiflora*, *Rumex acetosa*, *Stellaria holostea*, *Veronica chamaedrys* or *V. officinalis*.

Bryophyte layer often dense, but less diverse than W11b, with *Lophocolea bidentata*, *Plagiomnium undulatum* and *Rhytidiadelphus triquetrus*.

W11d *Stellaria holostea – Hypericum pulchrum* sub-community

Sub-community descriptions
W11a *Dryopteris dilatata* sub-community
The least heavily-grazed sub-community, typically forming closed-canopy high forests, where sessile oak is abundant and often co-dominant

with downy birch. Rowan may form a sub-canopy, and hazel is frequent. Hawthorn is preferential but rare. Often with an underscrub of *Dryopteris affinis* ssp. *borreri, D. dilatata, Lonicera periclymenum* and *Rubus fruticosus*. Grasses are less prominent than in the rest of the community and bryophytes are reduced owing to competition from tall herbs.

W11b *Blechnum spicant* sub-community
The most oceanic sub-community. Birches dominate, with occasional oak and scarce hazel. Grasses and small herbs (*Oxalis acetosella, Galium saxatile, Potentilla erecta*) are abundant, and ferns are often conspicuous. *Pteridium aquilinum* can dominate on deeper soil. *Blechnum spicant* is characteristic, and often luxuriant, with occasional *Oreopteris limbosperma* and *Athyrium filix-femina*, which may be lush in ravines. Bryophytes are abundant, particularly on leached, rocky sites. *Dicranum majus, Diplophyllum albicans, Hypnum cupressiforme, Isothecium myosuroides, Plagiochila spinulosa, Pleurozium schreberi, Polytrichum formosum* and *Rhytidiadelphus loreus* are all frequent, and there is often a good epiphytic flora, particularly on hazel bark.

On neutral to acidic soils in north-western Britain, W11b may be difficult to separate from W17c (*Anthoxanthum – Agrostis* sub-community of the *Quercus – Betula – Dicranum* woodland) particularly if grazing is heavy and sampling takes place late in the year, when *Hyacinthoides* has died back. The abundance and diversity of calcifuge bryophytes can help to separate the types, but there is a genuine convergence between them.

W11c *Anemone nemorosa* sub-community
Anemone nemorosa dominates in spring, followed by some of *Trientalis europaeus, Lathyrus linifolius, Luzula pilosa* and *Melampyrum pratense*. Ferns are scarce apart from *Pteridium aquilinum*, and although bryophytes are extensive, they are not as luxuriant as in W11a and W11b. This type is most common in north-east Scotland.

W11d *Stellaria holostea – Hypericum pulchrum* sub-community
The most mesophytic of the sub-communities. *Holcus mollis* is more common, and *Festuca rubra* and *Holcus lanatus* are preferential. There is an increase in the less acidophilous herb species, such as *Ajuga reptans, Cerastium fontanum, Hypericum pulchrum, Luzula multiflora, Stellaria holostea, Veronica chamaedrys* and *V. officinalis*. Amongst the bryophytes, *Eurhynchium praelongum, Lophocolea bidentata, Plagiomnium undulatum, Scleropodium purum, Rhytidiadelphus squarrosus* and *Thuidium tamariscinum* are common; whereas more acidophilous species are rare.

W12 *Fagus sylvatica – Mercurialis perennis* woodland

A community of free-draining base-rich calcareous soils (pH between 7 and 8) in the south-east lowlands of Britain, generally limited to the steeper drift-free faces of chalk escarpments. To the north-west, late frosts, low summer temperatures and heavier rainfall hinder beech dominance by their effects on mast production and regeneration, although beech woods can form well to the north-west of its natural range.

Beech is dominant throughout the community. Ash and sycamore are often present, often readily colonizing gaps. Pedunculate oak may occur but does not persist under deep shade. Whitebeam and yew are characteristic of the community, either as relicts of an early successional stand or persisting in areas where beech is not too tall. Yew is shade-tolerant and may persist as a shrub layer. Apart from this, the shrub layer is usually sparse, although a wide range of species may occur, including patches of hazel, hawthorn, field maple or holly.

Small gaps in the beech canopy may be dominated by ash, oak or sycamore but are often best treated as part of the beech community. Larger regeneration zones (more than about 75 m across) where beech is absent should be referred to the appropriate non-beech type.

The field layer consists of species characteristic of base-rich soils such as *Allium ursinum, Arum maculatum, Brachypodium sylvaticum, Circaea lutetiana, Galium odoratum, Melica uniflora, Mercurialis perennis, Mycelis muralis* and *Sanicula europaea. Hedera helix* can form a complete carpet and *Rubus fruticosus* is occasionally abundant, but where the shade is dense the field layer may be virtually absent. Plants of moist base-rich conditions such as *Ajuga reptans, Anemone nemorosa, Deschampsia cespitosa, Poa trivialis* or *Primula vulgaris* are rare.

Rubus fruticosus can also be abundant in W14 *(Fagus sylvatica – Rubus fruticosus* woodland), but in the latter *Mercurialis* and other calcicolous herbs and grasses tend to be rare, and more acidophilous species, such as *Lonicera periclymenum, Luzula pilosa, Oxalis acetosella* and *Pteridium aquilinum* are more common. However, even on base-rich soils, a field layer more typical of acid soils often occurs immediately around the base of beech trees, because of 'acid' run-off from the trunks.

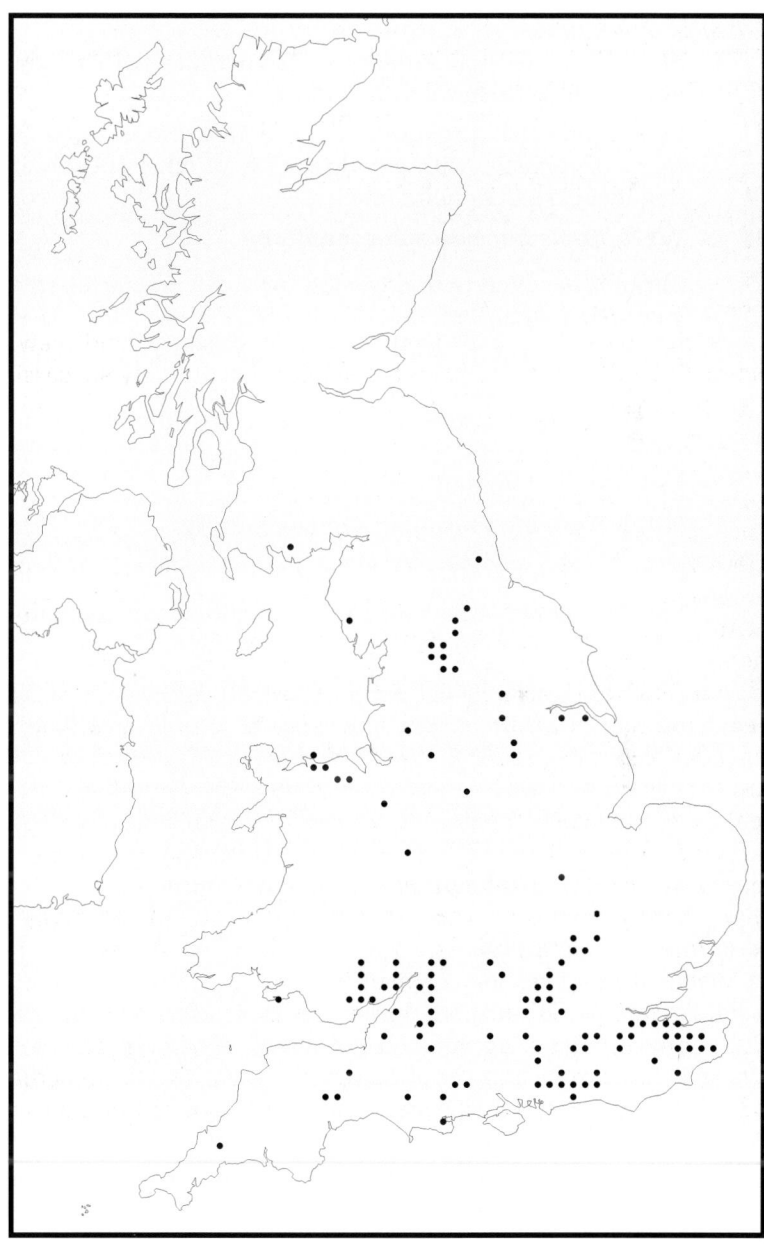

Figure 14 Distribution of W12 *Fagus sylvatica – Mercurialis perennis* woodland

Key to sub-communities:

Shrub layer with abundant yew, sometimes with whitebeam or box, but little hazel or hawthorn.

Field layer often a sparse mixture of *Mercurialis perennis* and *Rubus fruticosus*, with a wide variety of other species at low frequency. *Hedera helix* rare.

W12c *Taxus baccata* sub-community

OR Shrub layer often with dogwood, privet or wayfaring tree, but yew sparse or absent.

Field layer usually with *Hedera helix*, sometimes forming a ground carpet, with some of *Melica uniflora*, *Mycelis muralis*, *Poa nemoralis*, *Sanicula europaea* or *Tamus communis*. *Bromopsis ramosa*, *Clematis vitalba* and *Rumex sanguineus* may also be common.

W12b *Sanicula europaea* sub-community

OR Canopy often has abundant sycamore and ash as well as beech.

Shrub layer better developed than the other sub-communities, often with hazel or hawthorn. Yew sparse or absent.

Field layer usually with *Hedera helix*, sometimes forming a ground carpet, with some of *Arum maculatum*, *Brachypodium sylvaticum*, *Circaea lutetiana*, *Dryopteris filix-mas*, *Galium odoratum*, *Hyacinthoides non-scripta*, *Lamiastrum galeobdolon* or *Rubus fruticosus*. *Allium ursinum*, *Anemone nemorosa* or *Ranunculus ficaria* may be locally abundant in spring.

W12a *Mercurialis perennis* sub-community

Sub-community descriptions

W12a *Mercurialis perennis* sub-community
This sub-community occurs on deeper, moister soils than the others, usually on gently-sloping ground. Ash and sycamore are frequent associates of beech, with oak occasional. The understorey is patchy but better developed than in the other sub-communities. The field layer is dominated by *Mercurialis perennis*, and is consequently lush but species-poor, with a few taller herbs, such as *Brachypodium sylvaticum*, *Circaea lutetiana* or *Hyacinthoides non-scripta*, as well as *Hedera helix* and *Rubus fruticosus*.

W12b *Sanicula europaea* sub-community
This sub-community occurs on fairly steep slopes with shallow, well-drained soils. The canopy is overwhelmingly dominated by beech, and

the shrub layer is less extensive than in W12a. The drier soils limit the growth of *Mercurialis perennis*, and so the ground flora is more diverse. *Sanicula europaea* can be abundant, and *Mycelis muralis* is strongly preferential with *Brachypodium sylvaticum*, *Melica uniflora* and *Poa nemoralis* often giving a grassy appearance. A rich variety of orchids, including *Cephalanthera damasonium*, *Listera ovata* and *Neottia nidus-avis*, are also found in some stands.

W12c *Taxus baccata* sub-community

This sub-community is often found on still steeper, usually south-facing slopes with extremely thin and well-drained soils. Beech grows more slowly than in the other sub-communities, and species such as yew and whitebeam can keep pace, and become relatively common. The canopy height is consequently low, but casts a very deep shade, and so the shrub layer is very sparse, with some elder, hawthorn, privet, *Clematis vitalba* and occasionally box. The ground flora is often absent because of the shade, with scattered *Arum maculatum*, *Circaea lutetiana*, *Geum urbanum*, *Mycelis muralis* and *Melica uniflora* and some shade-tolerating mosses such as *Brachythecium rutabulum* and *Eurhynchium praelongum*. Other bryophytes, such as *Ctenidium molluscum*, *Encalypta streptocarpa* and *Homalothecium sericeum*, can be abundant in open areas.

W13a (*Sorbus aria* sub-community of *Taxus baccata* woodland) can be similar, but beech is generally rare in this sub-community, and never dominates the canopy.

W13 *Taxus baccata* woodland

A community of moderate to very steep, usually south-facing, limestone slopes carrying shallow dry rendzinas. It is almost common on the chalk of south-east England, on sites too dry for ash or beech woods and on limestone in northern England. While Rodwell (1991) treats these latter as part of the mixed deciduous woodland (W8), there is no reason why they should not be included in W13, and that has become the common practice.

Yew is the main canopy species, and rarely exceeds 10 m in height. Few other species occur, although ash, beech, pedunculate oak, sycamore and whitebeam may be present, usually as scattered trees. There is seldom a true shrub layer, but only scattered elder, holly or hawthorn, with box a rare associate. The dead woody remains of the preceding scrub, often of juniper, frequently occur. The field layer is extremely sparse, with just a patchy cover of *Mercurialis* with very occasional *Arum maculatum, Brachypodium sylvaticum, Fragaria vesca, Glechoma hederacea, Hedera helix, Rubus fruticosus, Urtica dioica* and *Viola riviniana*. The bryophyte cover is also very low.

Scattered stands of yew may occur in a mosaic with W8 (*Fraxinus excelsior – Acer campestre – Mercurialis perennis* woodland), W9 (*Fraxinus excelsior – Sorbus aucuparia – Mercurialis perennis* woodland) or W12 (*Fagus sylvatica – Mercurialis perennis* woodland). W12 and W13 may grade into one another where the frequency of beech is low.

Stands with an abundance of yew on acid substrates do not fit into the above scheme and if they prove widespread it may be necessary to create a new type.

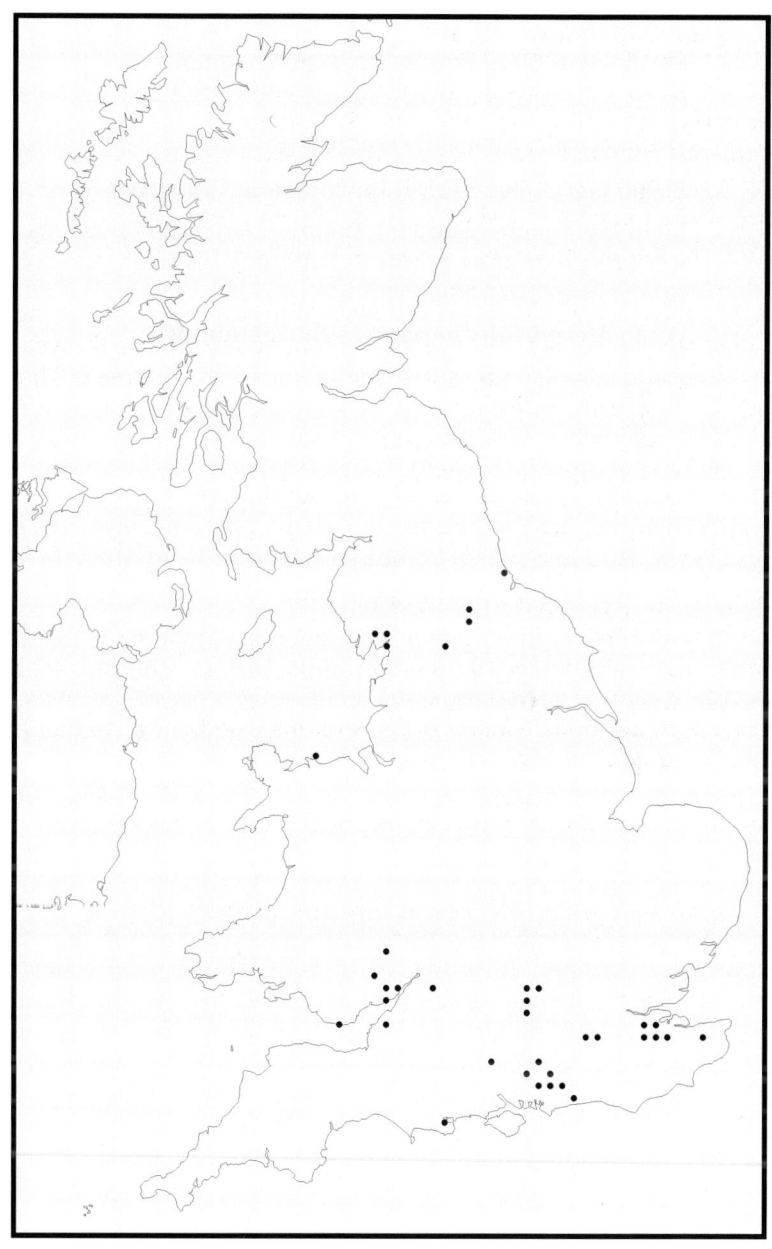

Figure 15 Distribution of W13 *Taxus baccata* woodland

Key to sub-communities

        Shrub layer often with whitebeam common.

        Field layer absent or very sparse.

        W13a *Sorbus aria* sub-community

OR   Shrub layer often with elder common. Whitebeam rare.

        Field layer more abundant, mainly *Mercurialis perennis* with some *Brachypodium sylvaticum*, *Fragaria vesca*, *Iris foetidissima*, *Rubus fruticosus*, *Viola* spp.

        W13b *Mercurialis perennis* sub-community

Sub-community descriptions

W13a *Sorbus aria* sub-community

Whitebeam is a frequent associate in the canopy, with some ash, beech, oak and sycamore. There is usually no shrub layer except occasional elder, although box can be locally important. The field layer is almost always absent, with just sparse bryophytes, bare soil and litter.

W13b *Mercurialis perennis* sub-community

Yew is the only species, in a rather more open canopy than W13a. There is often a sparse shrub layer with elder, and occasional spindle, dogwood and privet, with scrambling *Clematis vitalba* and *Tamus communis*. The field layer is a little less sparse than above (although most of the ground is bare of herbs), with constant *Mercurialis perennis*, and some *Brachypodium sylvaticum*, *Fragaria vesca*, *Iris foetidissima*, *Rubus fruticosus*, *Urtica dioica* or *Viola* spp.

W14 *Fagus sylvatica* – *Rubus fruticosus* woodland

A community confined to brown earth soils of low base status with moderate to slightly impeded drainage in south Britain, usually on superficial deposits (e.g. clay with flints) over the southern chalk. The pH is generally low (4-5) but leaching is limited.

    Stands tend to be dominated by beech which forms a closed, even-topped cover of very well-grown trees. There may be some structural complexity, relating to patterns of natural invasion or management, and in younger stands. Pollards are quite common, for example, in the New Forest.

    Pedunculate oak is the most common associate, with sessile oak on lighter soils, and stands can form transitions or mosaics with W10 (the edaphic equivalent of W14). Oak has a colonising advantage in younger woods, and beech takes over in older stands. Other species are scarce: there may be some birch, ash or sycamore in gaps, but less frequently than in W12. The shrub layer may be limited, but holly can be dense in oceanic areas. Yew, rowan, hawthorn, elder, hazel, privet and *Salix caprea* occur sporadically.

The field layer is sparse or absent where the canopy is dense, but *Rubus fruticosus* is usually the most common species and, where the shade is less intense, may form a continuous cover up to 1 m in height. Because of this cover, other species characteristic of W10 are poorly represented. *Pteridium aquilinum* and *Lonicera periclymenum* are frequent but not abundant, and *Hedera helix* and *Hyacinthoides non-scripta* are infrequent. *Oxalis acetosella* is characteristic before *Rubus fruticosus* attains dominance. Other species occurring as widely scattered individuals include *Deschampsia cespitosa*, *Dryopteris dilatata*, *D. filix-mas*, *Holcus mollis*, *Luzula pilosa*, *Melica uniflora*, *Milium effusum* and *Ruscus aculeatus*. *Galium odoratum* can be abundant but other calcicoles are rare. In gaps and around margins *Digitalis purpurea*, *Euphorbia amygdaloides*, *Rubus idaeus* and *Arctium minus* may be present, sometimes with *Epipactis helleborine* or, more rarely, *E. purpurata*. In the Wye Valley, *Luzula sylvatica* may dominate, as it does in many types of woodland in this area.

Bryophyte cover is generally poor, but may be more obvious where dense shade has excluded the herbs as, for example, around tree bases. Common species include *Atrichum undulatum*, *Dicranella heteromalla*, *Pseudotaxiphyllum elegans*, *Mnium hornum* and *Polytrichum formosum*. *Dicranum scoparium* and *Leucobryum glaucum* are rare, being more characteristic of W15.

No sub-communities.

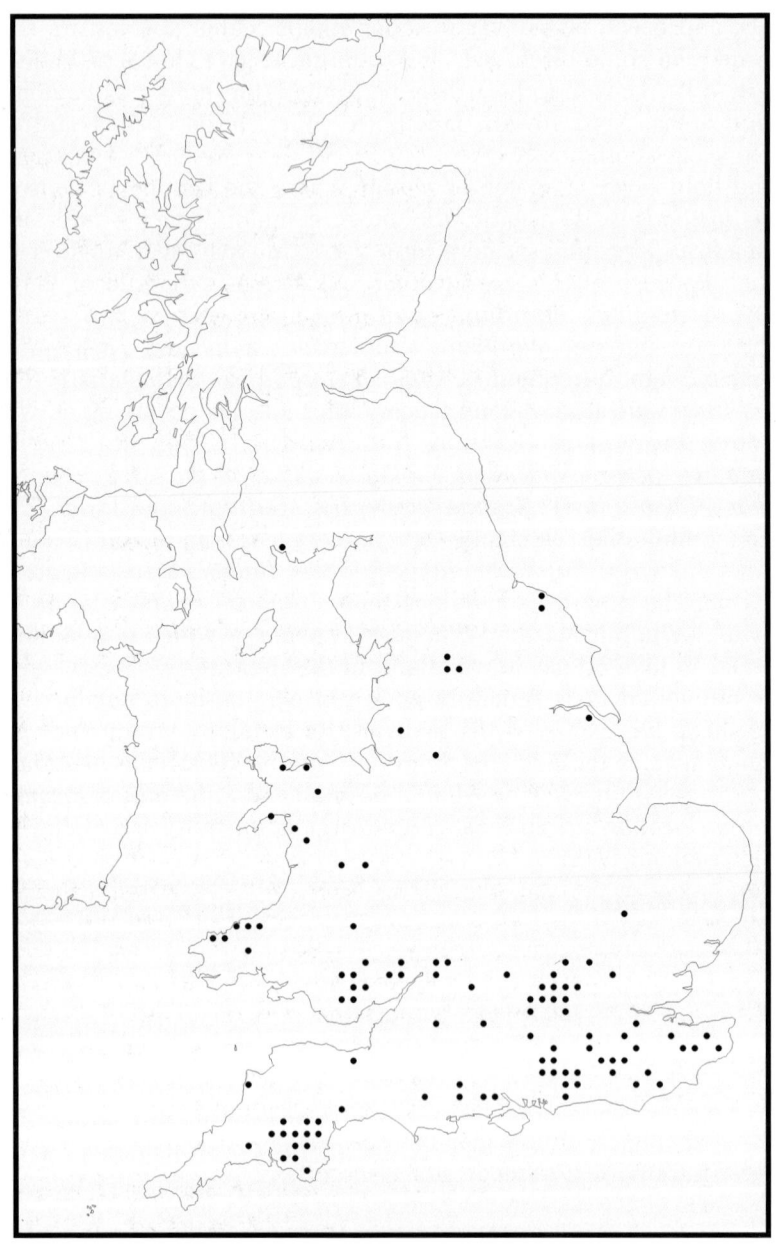

Figure 16 Distribution of W14 *Fagus sylvatica – Rubus fruticosus* woodland

W15 *Fagus sylvatica* – *Deschampsia flexuosa* woodland

A community of very base-poor infertile soil (pH <4) in the southern lowlands of Britain, usually on podzolic soils with mor humus and free to excessive drainage.

Pedunculate oak is the most common associate in the south, although sessile oak is locally frequent, and becomes more abundant in the north. Birch may be present in gaps, but sycamore, whitebeam and wild cherry are scarce and ash is absent. Stands usually have a high forest structure, and coppice is rare, but some stands have been treated as wood-pasture and retain old beech pollards (e.g. Burnham Beeches, Buckinghamshire). Small areas of regeneration dominated by oak or birch should be considered part of the beech type. Larger areas (more than 75 m across) where beech is absent should be classified as the appropriate non-beech community.

Because of the dense shade of the canopy, the shrub layer is often poor or absent. Holly is the main shrub-layer species, occasionally with some yew.

The ground is often bare of herbs, leaving expanses of litter and mor humus. Cover varies according to the density of the canopy, but most of the field layer is likely to be under small gaps or at the edge of stands. In addition beech is shallow-rooted and may exert considerable root competition for water. *Pteridium aquilinum* and *Deschampsia flexuosa* are the most frequent vascular plants and *Rubus fruticosus* is often present but less frequent than in W14. *Vaccinium myrtillus* and *Luzula sylvatica* occur in ungrazed areas. Other species which may be found include *Agrostis capillaris*, *Blechnum spicant*, *Dryopteris dilatata*, *Holcus mollis*, *Luzula pilosa*, *Melampyrum pratense*, *Oxalis acetosella*, and *Ruscus aculeatus*. The bryophyte layer is often distinctive with *Dicranella heteromalla*, *Dicranum scoparium*, *Hypnum cupressiforme*, *Pseudotaxiphyllum elegans*, *Leucobryum glaucum*, *Mnium hornum* and *Polytrichum formosum*. The community is renowned for its autumn fungi.

Most of the field layer development is likely to be under gaps or on the edges of stands. Differences between sub-communities are mainly related to the local light climate. Small areas of regeneration dominated by oak or birch within the beech stand should be considered as part of the beech type.

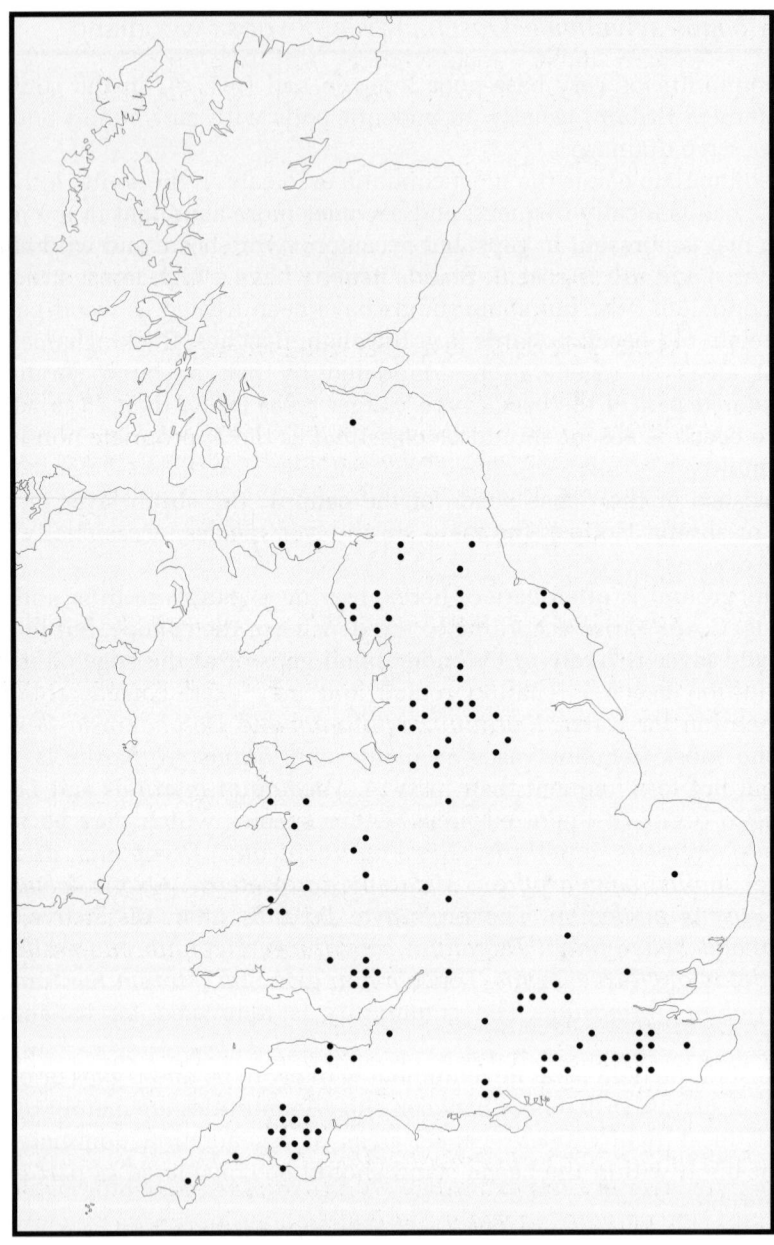

Figure 17 Distribution of W15 *Fagus sylvatica – Deschampsia flexuosa* woodland

Key to sub-communities

        Field layer species-poor with *Calluna vulgaris* and, sometimes, *Agrostis capillaris*, *Deschampsia flexuosa*, *Pteridium aquilinum* or *Vaccinium myrtillus*.

        W15d *Calluna vulgaris* sub-community

OR   Field layer with *Vaccinium myrtillus* (but no *Calluna vulgaris*), and often some of *Agrostis capillaris*, *Carex pilulifera*, *Deschampsia flexuosa*, *Hedera helix*, *Melampyrum pratense* and *Pteridium aquilinum*.

        W15c *Vaccinium myrtillus* sub-community

OR   Field layer with open or closed carpet of *Deschampsia flexuosa* (but no *Calluna vulgaris* and no more than very sparse *Vaccinium myrtillus*), sometimes with *Agrostis capillaris*, *Dryopteris dilatata*, *Holcus lanatus* or *Luzula pilosa*.

        W15b *Deschampsia flexuosa* sub-community

OR   Shrub layer very sparse.

        Field layer very sparse or absent. Sometimes patches of beech seedlings, *Dicranella heteromalla*, *Eurhynchium praelongum* or *Mnium hornum*. *Deschampsia flexuosa* usually absent.

        W15a *Fagus sylvatica* sub-community

Sub-community descriptions

W15a *Fagus sylvatica* sub-community
Beech is often the only species in a closed canopy, casting dense shade. There is often no shrub layer, although holly may be occasional. The field layer is negligible, often only a few tufts of bryophytes, such as *Dicranella heteromalla*, *Eurhynchium praelongum*, *Hypnum cupressiforme*, *Pseudotaxiphyllum elegans*, *Mnium hornum*.

W15b *Deschampsia flexuosa* sub-community
Beech is still dominant, but oak can be frequent to abundant, and holly is common in the shrub layer. The field layer comprises an open or closed carpet of *Deschampsia flexuosa*, with occasional *Pteridium aquilinum* and scattered *Agrostis capillaris*, *Holcus mollis*, *Luzula pilosa*. The bryophyte layer is more extensive, with *Dicranum scoparium*, *Lepidozia reptans*, *Leucobryum glaucum* and *Polytrichum formosum* as well as the above species.

W15c *Vaccinium myrtillus* sub-community
Oak is as common as beech, and birch is more frequent. There is often a dense holly shrub layer, and the ground flora is also richer. *Deschampsia flexuosa*, *Pteridium aquilinum* and *Vaccinium myrtillus* are constant, and *Carex pilulifera* and *Melampyrum pratense* are preferential, with patchy *Hedera helix*. Bryophytes are extensive, as in W15b, but *Leucobryum glaucum* is more frequent.

W15d *Calluna vulgaris* sub-community
Beech is the most abundant species in an open, discontinuous canopy, but oak and/or birch are often frequent with some holly. The field layer is extensive, and dominated by *Calluna vulgaris* and *Pteridium aquilinum*, with occasional *Agrostis capillaris*, *Deschampsia flexuosa*, *Rubus fruticosus* or *Vaccinium myrtillus*. Bryophytes are less frequent than in the other sub-communities.

W16 *Quercus* spp. – *Betula* spp. – *Deschampsia flexuosa* woodland

This community is confined to very acidic, oligotrophic soils (pH rarely above 4) in the lowlands and upland fringes. Soils are typically very free-draining, usually sandy and podzolic. Long-established stands occur as high-forest oak-coppice or in wood-pasture, but many stands are recent developments on heathland.

Both species of oak may be present, as may the hybrid, especially in Wales. Pedunculate oak tends to be prominent in the south and sessile oak in the north. Birch can be very abundant, and may dominate, especially in recently formed stands on old heathland, where self-sown pine may also be abundant. Other species are rarer, but beech, sweet chestnut, sycamore and whitebeam occur sporadically. The shrub layer is generally poor. Rowan and holly may be present, rowan being more frequent in the north, but hawthorn and hazel are very rare. The scarcity of hazel helps to separate W16 from W10. Alder buckthorn, elder and rhododendron may occur, the last sometimes forming dense thickets.

The field layer is generally species-poor. *Deschampsia flexuosa* and *Pteridium aquilinum* are the most consistent species. *Calluna vulgaris*, *Erica cinerea* and *Vaccinium myrtillus* may be frequent in ungrazed stands, particularly in the north-west, and *Agrostis capillaris* and *Anthoxanthum odoratum* are more common in grazed situations. *Luzula pilosa* can occur, and *L. sylvatica* may be locally abundant on steep slopes (cf W10, W11). Other species may be locally common, including

Blechnum spicant, Convallaria majalis, Ceratocapnos claviculata, Digitalis purpurea, Dryopteris dilatata, Galium saxatile, Hedera helix, Potentilla erecta, Rumex acetosella, Solidago virgaurea and *Teucrium scorodonia*. *Deschampsia cespitosa* and *Molinia caerulea* mark transitions to damper communities, such as W4 (*Betula pubescens – Molinia caerulea* woodland), but sedges are rare.

Dry soils and low atmospheric humidity limit the contribution of bryophytes in the east but they are more abundant to the north and west, particularly *Dicranella heteromalla, Dicranum scoparium, Hypnum cupressiforme, Pseudotaxiphyllum elegans, Leucobryum glaucum* and *Pleurozium schreberi*, although the diversity and abundance is less than in W17.

This community may be difficult to separate at times from the other 'oak' types (W10, W11, W17), but in general it has fewer 'Atlantic' bryophytes than W17, is less grassy than W11 (*Agrostis* spp., *Anthoxanthum odoratum* and *Holcus mollis* tend to be rare in W16) and has less *Hyacinthoides non-scripta, Lonicera periclymenum* and *Rubus fruticosus* than W10. W10d (the *Holcus lanatus* sub-community of the *Quercus robur – Pteridium aquilinum – Rubus fruticosus* woodland) in particular comes close to W16a.

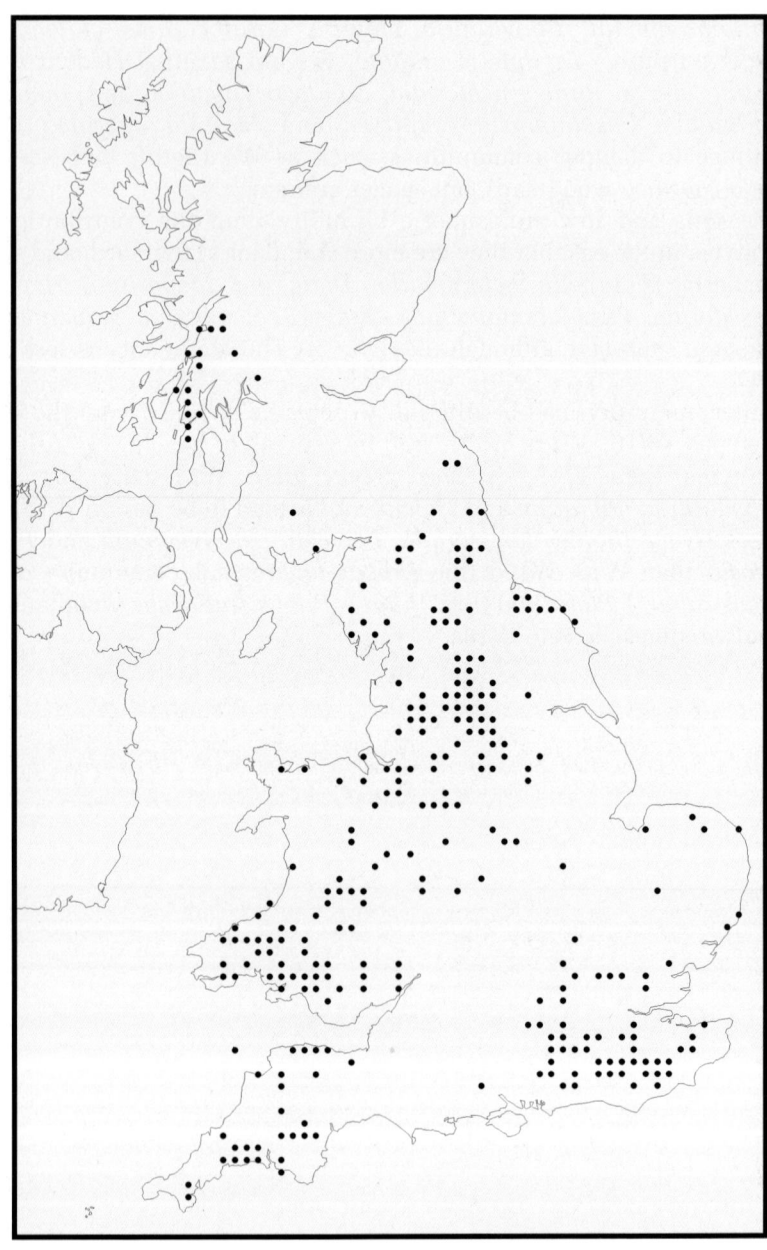

Figure 18 Distribution of W16 *Quercus* spp. – *Betula* spp. – *Deschampsia flexuosa* woodland

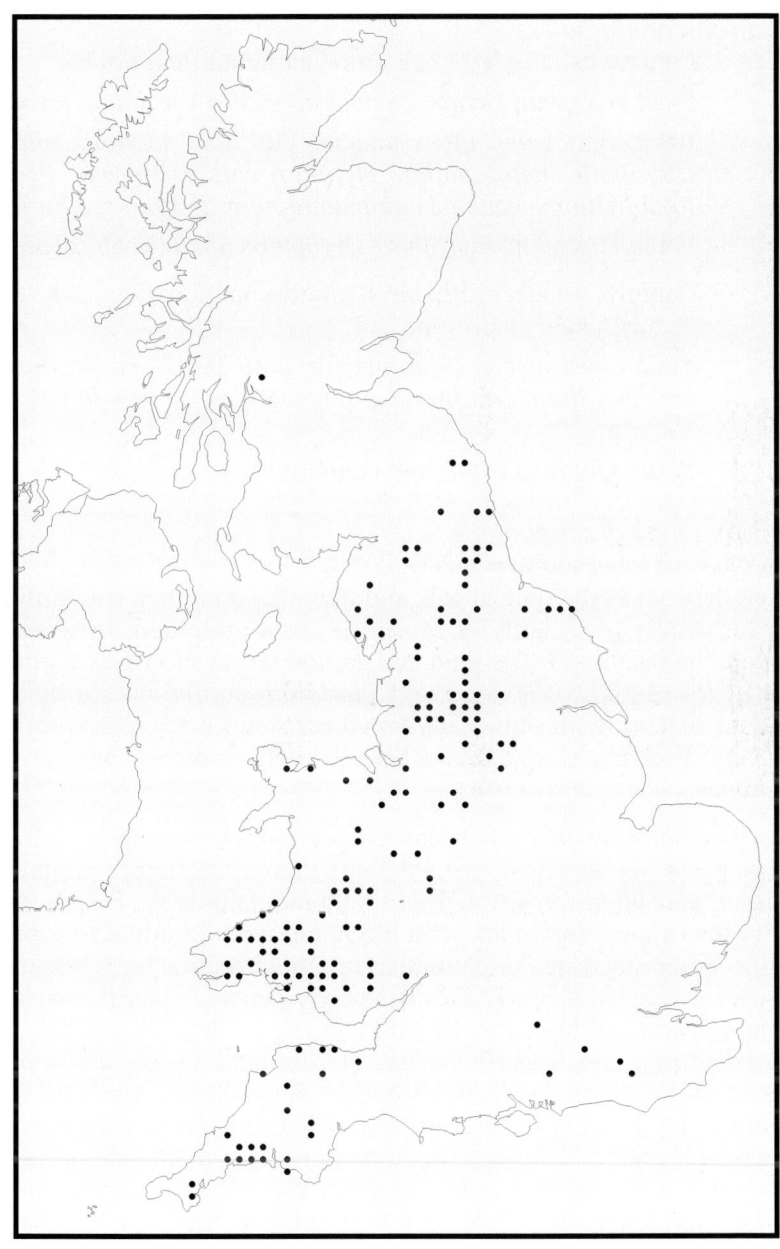

Figure 19 Distribution of W16b *Vaccinium myrtillus – Dryopteris dilatata* sub-community woodland

Key to sub-communities

Canopy usually with oak more abundant than birch.

Field layer with *Dryopteris dilatata* and/or *Vaccinium myrtillus*.

Bryophyte layer often sparse, but may include some of *Dicranella heteromalla, Hypnum cupressiforme, Pseudotaxiphyllum elegans, Lepidozia reptans* and *Mnium hornum*.

W16b *Vaccinium myrtillus – Dryopteris dilatata* sub-community

OR Canopy usually with birch more common than oak. Some stands may contain pine.

Field layer species-poor, usually with *Deschampsia flexuosa* and *Pteridium aquilinum* and sometimes *Holcus lanatus* and *Ulex* spp.

W16a *Quercus robur* sub-community

Sub-community descriptions

W16a *Quercus robur* sub-community
Pedunculate oak is the typical oak, and often the dominant tree, although birch (or locally pine, holly or rowan) is more prominent in secondary woodland on heathland. The field and ground layers show few distinctive features. *Deschampsia flexuosa* and *Pteridium aquilinum* are the most abundant species, with some *Calluna vulgaris* and *Erica cinerea* in open areas and *Vaccinium myrtillus* where rainfall is high. There are few bryophytes.

W16b *Vaccinium myrtillus – Dryopteris dilatata* sub-community
A more north-western type, characteristic of areas of higher rainfall and humidity, and often transitional with sub-montane W17. Sessile oak is usually the canopy dominant, with birch and rowan limited to gaps and margins. Some stands are of plantation origin with pine, larch and beech, and some are old coppices. The understorey is generally poorly developed, although rowan and holly can be common.

Deschampsia flexuosa, Pteridium aquilinum and ericoids are still common in the field layer, but there is more variety than in W16a. *Vaccinium myrtillus* is present on slopes but *Dryopteris dilatata* is the best preferential. The bryophyte layer can be rich in comparison to W16a, with some of *Dicranella heteromalla, Hypnum cupressiforme, Pseudotaxiphyllum elegans, Mnium hornum* and *Lepidozia reptans*.

W17 *Quercus petraea – Betula pubescens – Dicranum majus* woodland

A community of very acid, often thin soils in the cool, wet north-west of Britain, where there is a strong tendency for mor accumulation and high rainfall leads to strong leaching. Surface pH is usually below 4.

Sessile oak and/or downy birch usually dominate, although pedunculate oak is abundant in some localities (e.g. east Scotland and Dartmoor). Downy birch is particularly frequent to the north-west where oak is scarce. The canopy is often low and rather open, and in extreme cases may be very dwarfed. Rowan is the only other common tree species. It is often present as scattered individuals but can be locally abundant, and in the north-west often co-dominates with downy birch. Holly is often restricted by grazing, and ash and sycamore are restricted to enriched areas (tending to W9). Scattered beech and conifers, originating from planted stock, occur in places. The shrub layer is variable. Hazel is more abundant than in W16, but tends to be confined to deeper pockets of flushed soil. In the north-west it may form a scrubby canopy with downy birch and rowan.

The field layer is characterised by ericoid shrubs, *Pteridium aquilinum* and grasses. *Deschampsia flexuosa* is common with some *Agrostis capillaris*, *Anthoxanthum odoratum* and *Holcus mollis* on deeper soil, particularly in grazed woods, although the last three are more common in W11. *Agrostis vinealis*, *Festuca ovina* and *Molinia caerulea* may also occur. *Pteridium aquilinum* is abundant but confined to deeper soils and to areas which are not heavily shaded. *Vaccinium myrtillus* may be abundant, even in shaded situations, but is sensitive to grazing, as are *Calluna vulgaris* and *Erica cinerea*, which are also sensitive to shade. Small herbs are not abundant but *Galium saxatile*, *Luzula* spp., *Melampyrum pratense*, *Oxalis acetosella*, *Potentilla erecta*, *Solidago virgaurea* or *Teucrium scorodonia* often occur. *Hyacinthoides non-scripta* and *Anemone nemorosa* are infrequent. Ferns are often abundant, particularly *Blechnum spicant* but also *Dryopteris dilatata* and *Oreopteris limbosperma* with some *Athyrium filix-femina*, *Dryopteris affinis* ssp. *borreri*, *D. filix-mas*, *Gymnocarpium dryopteris* and *Polypodium vulgare*. *Dryopteris aemula*, *Hymenophyllum tunbrigense*, and *H. wilsonii* may occur on ledges.

The fern element attains its greatest abundance in ravines, as do bryophytes, which are particularly abundant in this community. Important species include *Dicranum majus*, *D. scoparium*, *Hylocomium splendens*, *Isothecium myosuroides*, *Plagiothecium undulatum*,

Key to sub-communities

Field layer may be sparse, depending on the terrain (often very rocky) and levels of grazing.

Bryophyte layer extremely rich, including *Diplophyllum albicans* and *Isothecium myosuroides* with four or more of *Bazzania trilobata, Campylopus flexuosus, Hypnum cupressiforme, Lepidozia reptans, Leucobryum glaucum, Plagiochila spinulosa, Scapania gracilis* or *Thuidium delicatulum*.

W17a *Isothecium myosuroides – Diplophyllum albicans* sub-community

OR Field layer grassy, usually with *Agrostis* spp., *Anthoxanthum odoratum, Galium saxatile* and *Holcus mollis*, and sometimes *Digitalis purpurea* and *Rubus fruticosus*.

Bryophyte layer often includes *Dicranella heteromalla, Eurhynchium praelongum, Lophocolea bidentata* and *Rhytidiadelphus squarrosus*.

W17c *Anthoxanthum odoratus – Agrostis capillaris* sub-community

OR Field layer heathy with *Calluna vulgaris* as well as *Vaccinium myrtillus. Erica cinerea, Luzula pilosa* and *Trientalis europaea* may also be present.

Bryophyte layer often includes *Scleropodium purum* and *Rhytidiadelphus triquetrus*.

W17d *Rhytidiadelphus triquetrus* sub-community

OR Field layer with none of the above combinations well developed. *Dryopteris dilatata* may be abundant, as well as the usual community constants.

W17b Typical sub-community

Sub-community descriptions

There is a general climatic trend from W17a (a very western sub-community, thriving in the high rainfall and humidity characteristic of the Atlantic region) to W17d (characteristic of eastern and central Scotland, where there is lower rainfall and humidity).

W17a *Isothecium myosuroides – Diplophyllum albicans* sub-community
This sub-community is the richest in bryophytes and tends to include the most notable western oceanic species. There are extensive mats, including

species typical of the thin humus which covers boulders and hangs over their steeper sides, and epiphytic cover extends far up tree trunks. Ferns, especially *Blechnum spicant*, are also often lush and *Hymenophyllum* spp. are occasional. Apart from the species mentioned above, *Dicranodontium denudatum*, *Heterocladium heteropterum*, *Hylocomium umbratum*, *Pseudotaxiphyllum elegans*, *Racomitrium heterostichum* and *Saccogyna viticulosa* are often among those present.

W17b Typical sub-community
Sessile oak is the most common canopy species, with downy birch only occasional. Usually less heavily grazed than W17a, and there is often an understorey of hazel common, frequently with holly and rowan. Ericoid shrubs, *Pteridium aquilinum* and *Dryopteris dilatata* are usually abundant, and small herbs uncommon. Bryophytes are abundant and Atlantic and epiphytic species may still be common, but the species characteristic of thin humus, which typify W17a are absent.

W17c *Anthoxanthum odoratum* — *Agrostis capillaris* sub-community
A more heavily grazed type, with only a very patchy hazel understorey. Ericoid shrubs are uncommon, but grasses are abundant. *Agrostis capillaris*, *Anthoxanthum odoratum*, *Galium saxatile* and *Holcus mollis* are constant, with frequent *Pteridium* and small herbs, especially *Oxalis acetosella*. *Digitalis purpurea* and *Rumex acetosa* are weakly preferential, and *Rubus fruticosus* occasional. Bryophytes are scarcer than in the other sub-communities, but the larger species, *Hylocomium splendens*, *Plagiothecium undulatum* and *Rhytidiadelphus loreus*, are usually present.

This is the closest sub-community to W11 (*Quercus petraea* – *Betula pubescens* – *Oxalis acetosella* woodland). However, in W17 the large bryophytes are more frequent and abundant, while *Dryopteris* spp., *Hyacinthoides non-scripta* and *Rubus fruticosus* are more common in W11.

W17d *Rhytidiadelphus triquetrus* sub-community
Pedunculate oak often replaces sessile oak, in what is often a single-layered canopy, but downy birch and rowan dominate. Ericoid shrubs are common, and *Calluna vulgaris* is preferential. *Pteridium aquilinum* is often dense and small herbs are frequent, including *Galium saxatile*, *Melampyrum pratense*, *Oxalis acetosella*, *Potentilla erecta* and *Viola riviniana*, and *Trientalis europaea* is an occasional preferential. Bryophytes are common, with *Rhytidiadelphus triquetrus* and *Scleropodium purum* frequent but Atlantic species rare.

W18 *Pinus sylvestris* – *Hylocomium splendens* woodland

A community of strongly leached, lime-free, podzolic soils in the central and north-western highlands of Scotland. Variation in composition is generally related to the density and age of the pine canopy, but climate, soils and the incidence of browsing, grazing and burning are also important in defining sub-communities.

Scots pine is always the most abundant tree. The canopy is usually low (13-15 m, rarely 20 m) and often open, particularly in the west (an arbitrary lower limit of 25% cover separates W18 from ericoid heath), with denser stands in eastern Scotland. Pine tends to occur as a mosaic of well-segregated age-classes and the structural variation is often reflected in the field layer. Birch is the next most common tree, downy birch in the west or silver birch in the east, and rowan may be locally common. Where these are all abundant they represent a transition to W17 or W11, and mosaics with these types are common. Juniper is also sometimes present as scattered bushes, but can form patches excluding pine, forming mosaics of W18 and W19.

Deschampsia flexuosa is usually present, and is abundant in grazed situations or under dense shade where ericoid shrubs are reduced. *Calluna vulgaris*, *Vaccinium myrtillus* and *V. vitis-idaea* are more frequent than in other woodland types, but their abundance is variable. *Calluna vulgaris* is sensitive to shade, and so is more prominent under open canopies. *Calluna* and *V. myrtillus* are sensitive to grazing, and where this is heavy *V. vitis-idaea* predominates. Other sub-shrubs which may be prominent include *Empetrum nigrum*, *Erica cinerea* and *E. tetralix*. *Pteridium aquilinum* is usually present but less common than in other acidic woods, despite the generally open canopy. *Molinia caerulea* occurs especially in western stands and in transitions to mires. *Agrostis capillaris*, *A. vinealis*, *Anthoxanthum odoratum* and *Festuca ovina* can be prominent where grazing is heavy. Other herbaceous species are scarce but may include *Melampyrum pratense*, *Potentilla erecta*, *Trientalis europaea*, *Luzula pilosa*, *Oxalis acetosella* and *Galium saxatile*. Herbs with a strong continental northern distribution are a characteristic of these woods: *Goodyera repens* is most abundant of these with less frequently *Listera cordata*, *Pyrola minor*, *P. media*, *P. rotundifolia*, *Moneses uniflora*, *Orthilia secunda* and *Linnaea borealis*.

Bryophytes can be a prominent component of the field layer, particularly *Dicranum scoparium*, *Hylocomium splendens*, *Plagiothecium undulatum*, *Pleurozium schreberi* and *Rhytidiadelphus loreus*, which are common in other north-western acidic types, and *Ptilium crista-castrensis*, which is more restricted and preferential to W18. Lichens, particularly *Cladonia* spp., are often scattered through the bryophyte mat.

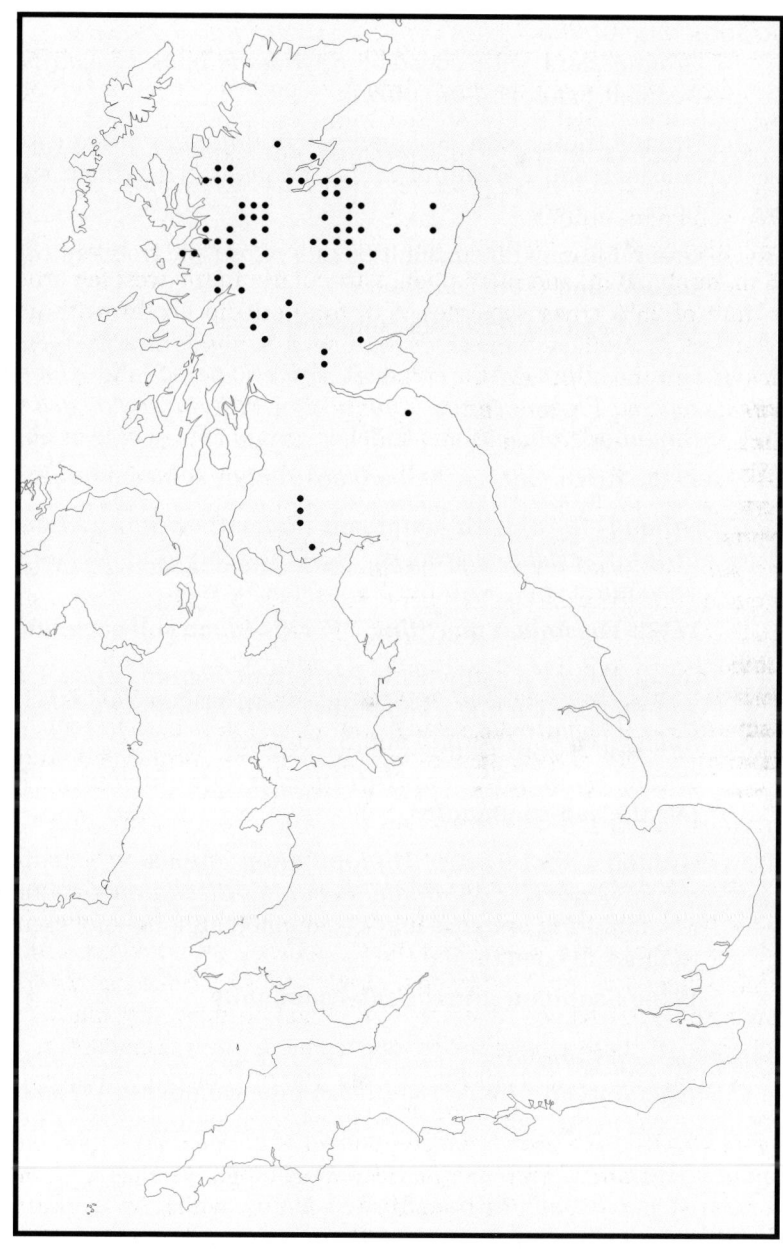

Figure 21 Distribution of W18 *Pinus sylvestris – Hylocomium splendens* woodland

Key to sub-communities

1	Ground flora with constant *Rhytidiadelphus triquetrus* and frequent *Scleropodium purum*.	→ 2
OR	Ground flora with *Sphagnum capillifolium/quinquefarium* constant and *Dicranum majus* frequent. The above species rare.	→ 3
2	Ground flora with constant *Luzula pilosa* and frequent *Galium saxatile* and *Oxalis acetosella*.	
	W18c *Luzula pilosa* sub-community	
OR	Ground flora with constant *Erica cinerea* and *Goodyera repens*. *Plagiothecium undulatum*, *Ptilium crista-castrensis*, *Rhytidiadelphus loreus* and *Vaccinium* spp. sparse or absent.	
	W18a *Erica cinerea – Goodyera repens* sub-community	
OR	Ground flora with abundant *Plagiothecium undulatum*, *Ptilium crista-castrensis*, *Rhytidiadelphus loreus* and *Vaccinium* spp., with the above species rare.	
	W18b *Vaccinium myrtillus – V. vitis-idaea* sub-community	
3	Ground flora usually with *Empetrum nigrum*, *Erica tetralix* and *Molinia caerulea*, sometimes with *Leucobryum glaucum*, *Sphagnum girgensohnii* and *S. russowii*.	
	W18d *Sphagnum capillifolium/quinquefarium – Erica tetralix* sub-community	
OR	Ground flora with *Diplophyllum albicans*, *Scapania gracilis* and *Thuidium tamariscinum*. *Blechnum spicant* and *Melampyrum pratense* may be common although not confined to this sub-community.	
	W18e *Scapania gracilis* sub-community	

Sub-community descriptions
There is a general trend from W18a (dense pine canopy) to W18e (open pine canopy, associated with a greater degree of leaching in the soil).

W18a *Erica cinerea – Goodyera repens* sub-community
Erica cinerea is constant, although often at low cover, with *Goodyera repens*. Other ericoids are sparse, and *Deschampsia flexuosa* is often the most common vascular plant. Bryophytes are generally the most prominent element of the field layer. *Rhytidiadelphus triquetrus* and *Hylocomium*

splendens are usually the most frequent, with *Dicranum scoparium*, *Hypnum jutlandicum*, *Lophocolea bidentata*, *Pleurozium schreberi* and *Scleropodium purum*.

W18b *Vaccinium myrtillus – V. vitis-idaea* sub-community
The canopy is typically a little more open than in W18a, and sub-shrubs are much more dense, particularly *Vaccinium* spp., often with some *Calluna vulgaris* and *Empetrum nigrum*. Scattered small herbs occur, particularly in less dense areas, e.g. *Melampyrum pratense* and *Goodyera repens*. Bryophytes are often rich, and often more diverse than in W18a, although they can be obscured by ericoids. *Rhytidiadelphus loreus*, *Ptilium crista-castrensis* and *Plagiothecium undulatum* can be common, as well as the species listed above.

W18c *Luzula pilosa* sub-community
This sub-community is quite similar to W18b, with a field layer dominated by ericoids and a rich bryophyte carpet. Small herbs, however, are much more frequent. *Luzula pilosa* is constant and *Galium saxatile* and *Oxalis acetosella* are common. *Blechnum spicant* can also be common, particularly in the east.

W18d *Sphagnum capillifolium – Erica tetralix* sub-community
The canopy is typically much less dense here, often with some birch and rowan and, in more southerly stands, a little juniper. The topography is generally uneven, with bryophyte-covered pine stumps and boulders, and hollows where the woodland gives way to more boggy vegetation. Amongst the sub-shrubs, *Calluna vulgaris* dominates, although *Vaccinium vitis-idaea* and *Empetrum nigrum* are very frequent, and *Erica tetralix* is common. The really distinctive feature of the ground layer is the prominence of deep sphagnum tussocks.

W18e *Scapania gracilis* sub-community
Very similar to W18d but usually lacking *Erica tetralix*, and with a much richer, more oceanic bryophyte layer. *Diplophyllum albicans*, *Scapania gracilis* and *Thuidium tamariscinum* are all frequent.

4 References

Blockeel, T L, & Long, D G (1998) *A check-list and census catalogue of British and Irish bryophytes*. British Bryological Society, Cardiff.
http://rbg-web2.rbge.org.uk/bbs/Resources/uklist.htm

Cooke, R J and Kirby, K J (1994) The use of a new woodland classification in surveys for nature conservation purposes in England and Wales. *Arboricultural Journal* **18**, 167-186.

Coppins, B J (2002) *Checklist of lichens of Great Britain and Ireland*. British Lichen Society, London. http://users.argonet.co.uk/users/jmgray/numlst.htm

Dony, J D, Jury, S L & Perring, F (1986) *English names of wild flowers*. 2nd edn. Botanical Society of the British Isles, Reading.

Goldberg, E (ed.) (2003) National Vegetation Classification – ten years' experience using the woodland section. *JNCC Report*, No. 335
www.jncc.gov.uk/publications/jncc335/default.htm

Hall, J E (1996) NVC database for woodlands. *English Nature Research Reports*, No. 181.

Hall, J E (1997) An analysis of National Vegetation Classification survey data. *JNCC Report*, No. 272.

Kirby, K J, Saunders, G R and Whitbread, A M (1991) The National Vegetation Classification in nature conservation surveys – a guide to the use of the woodland section. *British Wildlife* **3**, 70-80.

Palmer, M (1992) Trial of MATCH and TABLEFIT computer programs for placing survey data within the National Vegetation Classification. *JNCC Report*, No. 20.

Peterken, G P (1981) *Woodland conservation and management*. Chapman and Hall, London.

Rodwell, J S (ed.) (1991) *British Plant Communities. Volume 1. Woodlands and scrub*. Cambridge University Press, Cambridge.

Rodwell, J S (ed.) (2000) *British Plant Communities. Volume 5. Maritime communities and vegetation of open habitats*. Cambridge University Press, Cambridge

Rodwell, J S, Dring, J C, Averis, A B G, Proctor, M C F, Malloch, A J C, Schaminée, J N J & Dargie, T C D (2000) Review of coverage of the National Vegetation Classification. *JNCC Report*, No. 302.
www.jncc.gov.uk/Publications/JNCC302/INTR302.htm

Stace, C (ed.) (1997) *New flora of the British Isles*. 2nd edn. Cambridge University Press, Cambridge

Strachan, I M & Jackson, D (2003) Review and development of the National Vegetation Classification: stability and change. *In*: National Vegetation Classification – ten years' experience using the woodland section, ed. by E. Goldberg, 87-92. *JNCC Report*, No. 335

Whitbread, A M and Kirby, K J (1992) *Summary of National Vegetation Classification woodland descriptions*. Joint Nature Conservation Committee, Peterborough, (UK Nature Conservation No. 4).

5 Further reading

Hall, J E and Kirby, K J (1998) The relationship between Biodiversity Action Plan Priority and Broad Woodland Habitat Types and other woodland classifications. *JNCC Report*, No. 288. www.jncc.gov.uk/habitats/jncc288/default.htm

Kirby, K J (1988) *A woodland survey handbook.* Nature Conservancy Council, Peterborough. (Research and survey in nature conservation No. 11).

Appendix I: Relationships between different woodland classification systems

BAP Priority Habitat	Forestry Commission Guide Type	CORINE
Lowland beech and yew woodland	1. Lowland acid beech and oak woods 2. Lowland beech – ash woods	42.A71 41.13 41.12 (41.16)
Lowland mixed deciduous woodland*	1. Lowland acid beech and oak woods 3. Lowland mixed broadleaved woods	41.23, 41.32 (41.24) (41.51) 41.52
Upland mixed ashwoods	4. Upland mixed ashwoods	41.31, 41.32, 41.41 42.A71 (62.3)
Upland oakwood	5. Upland oakwoods	41.53, 41.52
Upland birchwoods*	6. Upland birchwoods	41.53, 41.52
Native pine woodlands	7. Native pinewoods	42.51 44.A2
Wet woodland	8. Wet woodland	44.A1 44.31 44.13 44.92
Lowland wood-pastures and parkland[4]	Referred to particularly in Lowland acid beech and oak woods guide (1), but no real equivalent	84.5 (but not a good equivalent)

Habitats Directive Annex I Type	NVC Type	Stand Types
Taxus baccata woods of the British Isles *Asperulo–Fagetum* beech forests Atlantic acidophilous beech forests with *Ilex* and sometimes also *Taxus* in the shrublayer	W13 W12 W14, W15	No direct equivalent 8C 8A, 8B, 8D, 8E
(*Tilio–Acerion* forests of slopes, screes and ravines) (Sub-Atlantic and medio-European oak or oak-hornbeam forests (Old acidophilous oak woods with *Quercus robur* on sandy plains)	W8a-d (e-g) W10a-d (e),[1] W16	1(A), B, 2, 3A, B, 4, 7C, 9, 10 5, 6C, D, 9, 12
Tilio–Acerion forests of slopes, screes and ravines) *Taxus baccata* woods of the British Isles Limestone pavements	(W7c) W8(a-c)[2] d-g , W9 W13	1A, C, D, 3C, D, 4C, 7D No direct equivalent
Old sessile oak woods with *Ilex* and *Blechnum* in the British Isles	W10e, W11, W16b, W17	6A, 6B (8A, 8B)
Old sessile oak woods with *Ilex* and *Blechnum* in the British Isles	W10e, W11, W17[3] W4a, b	Mainly stand group 12
Caledonian forest Bog woodland	W18, (W19) W4(a), b, c	11
Bog woodland Alluvial forests with *Alnus glutinosa* and *Fraxinus excelsior*	W4(a, b), c W5-W7 (W8) W1, W2, W3	(12) 7A, B, D, E No equivalent
Includes examples of Atlantic acidophilous beech forests with *Ilex* and sometimes also *Taxus* in the shrublayer. Old acidophilous oak woods with *Quercus robur* on sandy plains	W14, W15 W10, W16	Group 8 Group 6

*These types were proposed as new Priority Habitat types to the UK Targets Group. A decision on their status is awaited.

1 W8e-g and W10e are generally considered as part of upland mixed ash and upland oak respectively. However, they can occur in the lowlands, particularly in south-east England, and are then considered as part of lowland mixed broadleaves.

2 W8a–c are generally considered as part of lowland mixed broadleaves. However, they can occur in upland situation in a mosaic with upland types, and should then be considered as part of upland mixed ash.

3 Upland birchwoods and upland oakwoods cannot be separated on the grounds of NVC type. Both are largely composed of W10e, W11 and W17. Upland birchwoods occur in Scotland, largely beyond the altitudinal and latitudinal range of oak. The ground flora and bryophyte layers are very similar.

4 There is no good equivalent to this type in the other classification systems used. It is not a vegetation type but a management system and, theoretically at least, any NVC community or Stand type could be managed as wood pasture. Community types listed are those most frequently found in wood pasture and parkland.

Appendix II: Floristic tables

For each community a floristic table has been produced (Rodwell 1991). The Table below shows the floristic table for W2 *Salix cinerea – Betula pubescens – Phragmites australis* woodland. The species in these tables are grouped according to their occurrence in the canopy, shrub layer or field layer, including bryophytes. For each of these layers the first group of species is those which characterise the community. Where there are sub-communities, the species within each layer that separate the sub-communities follow those characterising the community. Then comes a tail of other species that were recorded in the samples. Some of these may be quite abundant. The explanation of the groupings is given after the table.

	a	*b*	2
Betula pubescens	III (3-8)	V (6-9)	I (3-9)
Salix cinerea	III (3-9)	IV (2-6)	I (2-9)
Frangula alnus	I (1-9)	II (4-6)	I (1-9)
Quercus robur	I (6)	I (1)	I (1-6)
Salix aurita	I (4)	I (3)	I (3-4)
Alnus glutinosa	III (4-10)	I (1-4)	II (1-10)
Fraxinus excelsior	II (3-6)	I (1)	I (1-6)
Crataegus monogyna	II (2-5)	I (1-5)	I (1-5)
Viburnum opulus	II (1-6)		I (1-6)
Salix fragilis	I (5-8)		I (5-8)
Rhamnus catharticus	I (3-5)		I (3-5)
Alnus glutinosa sapling	I (3-4)	I (5)	I (3-5)
Fraxinus excelsior sapling	I (1-5)	I (1-7)	I (1-7)
Betula pendula sapling		II (4-6)	I (4-6)
Phragmites australis	V (2-9)	IV (2-8)	IV (2-9)
Filipendula ulmaria	IV (1-7)	II (3-4)	III (1-7)
Brachythecium rutabulum	IV (2-7)	II (2)	III (2-7)
Urtica dioica	III (2-7)	I (1)	II (1-7)

	a	b	2
Eupatorium cannabinum	III (3-6)	I (1-3)	II (1-6)
Plagiomnium undulatum	II (2-5)	I (2)	I (2-5)
Galium palustre	II (1-4)	I (1-3)	I (1-4)
Cirsium palustre	II (1-4)	I (2)	I (1-4)
Carex acutiformis	II (2-9)		I (2-9)
Epilobium hirsutum	II (1-4)		I (1-4)
Galium aparine	II (2-5)		I (205)
Angelica sylvestris	II (1-4)		I (1-4)
Mentha aquatica	II (3-5)		I (3-5)
Solanum dulcamara	II (1-5)		I (1-5)
Caltha palustris	I (2-4)		I (2-4)
Phalaris arundinacea	I (3-7)		I (3-7)
Stellaria media	I (3)		I (3)
Lycopus europaeus	I (2-4)		I (2-4)
Calystegia sepium	I (4-5)		I (4-5)
Carex acuta	I (4-8)		I (4-8)
Epilobium palustre	I (2-3)		I (2-3)
Hedera helix	I (1-5)		I (1-5)
Plagiothecium nemorale (= *P. sylvaticum*)	I (2-3)		I (2-3)
Leptodictyum riparium (= *Amblystegium riparium*)	I (2-3)		I (2-3)
Symphytum officinale	I (2)		I (2)
Fraxinus excelsior seedling	I (2-3)		I (2-3)
Tamus communis	I (3-4)		I (3-4)
Amblystegium serpens	I (2-3)		I (2-3)
Geranium robertianum	I (1-4)		I (1-4)
Deschampsia cespitosa	I (3-4)		I (3-4)
Glechoma hederacea	I (2-4)		I (2-4)
Humulus lupulus	I (3-5)		I (3-5)
Sphagnum squarrosum	I (4)	V (2-7)	III (2-7)
Sphagnum fimbriatum	I (5)	V (4-7)	III (4-7)
Sphagnum fallax (= *S. recurvum*)	I (3)	IV (2-6)	III (2-6)

Sphagnum palustre		IV (3-8)	III (3-8)
Lonicera periclymenum	I (3-5)	III (1-6)	II (1-6)
Mnium hornum	I (1-3)	III (1-3)	II (1-3)
Plagiothecium denticulatum	I (1)	III (1-3)	II (1-3)
Holcus lanatus	I (1-5)	III (2-5)	II (1-5)
Juncus effusus	I (2-3)	III (2-5)	II (2-5)
Dryopteris carthusiana	I (3-4)	II (1-3)	I (1-4)
Hydrocotyle vulgaris	I (4)	II (3-4)	I (3-4)
Molinia caerulea	I (5)	II (2-6)	I (2-6)
Potentilla erecta	I (3)	II (2-4)	I (2-4)
Calypogeia fissa	I (2-3)	II (1-2)	I (1-3)
Myrica gale		II (5)	I (5)
Dryopteris cristata		II (1-3)	I (1-3)
Aulacomnium palustre		II (2)	I (2)
Rhizomnium pseudopunctatum		II (2-3)	I (2-3)
Agrostis canina		II (3-7)	I (3-7)
Agrostis stolonifera		II (1-4)	I (1-4)
Carex vesicaria		I (4-6)	I (4-6)
Calliergon giganteum		I (3-7)	I (3-7)
Deschampsia flexuosa		I (4)	I (4)
Sphagnum subnitens		I (1-4)	I (1-4)
Thelypteris phegopteris		I (3-8)	I (3-8)
Carex nigra		I (2)	I (2)
Oreopteris (= Thelypteris) limbosperma		I (3-4)	I (3-4)
Orthodontium lineare		I (2-3)	I (2-3)
Quercus petraea seedling		I (1)	I (1)
Menyanthes trifoliata		I (2-3)	I (2-3)
Eurhynchium praelongum	III (2-6)	III (2-5)	III (2-6)
Dryopteris dilatata	II (1-4)	II (2)	II (1-4)
Poa trivialis	II (2-7)	II (3-4)	II (2-7)
Rubus fruticosus agg.	II (2-8)	II (2-4)	II (2-8)

	a	b	2
Thelypteris palustris	II (4-5)	II (1-5)	II (1-5)
Ajuga reptans	I (3-4)	II (2-4)	I (2-4)
Lotus pedunculatus (= *L. uliginosus*)	I (3-4)	II (3)	I (3-4)
Rosa canina agg.	I (2-4)	II (2-4)	I (2-4)
Athyrium filix-femina	I (1-4)	I (2-4)	I (1-4)
Berula erecta	I (4)	I (3)	I (3-4)
Carex paniculata	I (3-7)	I (1)	I (1-7)
Carex remota	I (1-3)	I (1-3)	I (1-3)
Cladium mariscus	I (2-4)	I (1-3)	I (1-4)
Equisetum palustre	I (1-4)	I (2-3)	I (1-4)
Peucedanum palustre	I (3)	I (1-3)	I (1-3)
Lythrum salicaria	I (3-4)	I (1-3)	I (1-4)
Lysimachia vulgaris	I (2)	I (1-3)	I (1-3)
Juncus subnodulosus	I (4)	I (1-3)	I (1-4)
Glyceria maxima	I (1-2)	I (3)	I (1-3)
Calamagrostis canescens	I (2-3)	I (1-4)	I (1-4)
Rubus idaeus	I (4)	I (2)	I (2-4)
Pellia epiphylla	I (2)	I (2)	I (2)
Lophocolea heterophylla	I (1-3)	I (2)	I (1-3)
Scutellaria galericulata	I (3)	I (2)	I (2-3)
Lophocolea bidentata	I (4)	I (2-3)	I (2-4)
Valeriana officinalis	I (2-4)	I (1)	I (1-4)
Calliergonella cuspidata (= *Calliergon cuspidatum*)	I (4-5)	I (1-3)	I (1-5)
Atrichum undulatum	I (3)	I (1)	I (1-3)
Campylopus flexuosus (= *C. paradoxus*)	I (3)	I (2)	I (2-3)
Hypnum cupressiforme	I (3)	I (2)	I (2-3)
Pohlia nutans	I (4)	I (3)	I (3-4)
Number of samples	33	11	44
Number of species/sample	18 (7-27)	23 (15-30)	19 (7-30)

a: *Alnus glutinosa* – *Filipendula ulmaria* sub-community
b: *Sphagnum* sub-community
2: *Salix cinerea* – *Betula pubescens* – *Phragmites australis* woodland (total)

In the Table, the species are grouped as follows:

Betula pubescens – Salix aurita: canopy species found more or less equally in both sub-communities;

Alnus glutinosa – Rhamnus catharticus: canopy species more common in W2a than in W2b;

Alnus glutinosa sapling *– Fraxinus excelsior* sapling: shrub layer species found more-or-less equally in both sub-communities;

Betula pendula sapling: shrub layer species slightly more common in W2b than in W2a;

Phragmites australis: field layer species which is a constant for the community;

Filipendula ulmaria – Humulus lupulus: field layer species more common in W2a than in W2b;

Sphagnum squarrosum – Menyanthes trifoliata: field layer species more common in W2a than in W2b;

Eurhynchium praelongum – Pohlia nutans: field layer species common to both sub-communities.

Species which are distinctly more frequent in one or (where there are more than two) more of the sub-communities than the others are referred to as preferential. For example, the group *Filipendula ulmaria* to *Humulus lupulus* is particularly characteristic of W2a, although some of the species also occur in W2b. Even quite uncommon plants can be good preferentials, such as *Caltha palustris*; it is not often found in W2 but, when it is, it is generally in W2a.

For each species frequency and abundance values are given in the table. Frequency refers to how often a plant is found in a series of vegetation samples, irrespective or how much of that species is present in each sample. This is summarised in the table as classes denoted by the Roman numerals I to V, as follows:

1-20% frequency (up to one sample in five)	I	rare
21-40% frequency	II	occasional
41-60% frequency	III	frequent
61-80% frequency	IV	constant
81-100% frequency	V	constant

Constant species are those recorded in classes IV and V. Constant does not mean that the species occurred in every sample. A constant species could occur in only 61% of samples (and therefore be absent from 39% of samples).

Abundance describes how much of a species is present in a sample, irrespective of how frequent or rare it is among the samples. This is summarised in the table as bracketed numbers for the Domin ranges, and denoted in the text using terms such as dominant, abundant, plentiful and sparse.

Appendix III: Key bryophytes and their frequency in different woodland types (as recorded in the NVC tables)

	W1	W2	W3	W4	W5	W6
Atrichum undulatum		I			I	
Brachythecium rivulare						
Brachythecium rutabulum	I	III	II	I	IV	II
Calliergonella cuspidata (= *Calliergon cuspidatum*)	I	I	V	I	II	
Cirriphyllum piliferum						
Climacium dendroides			III			
Dicranella heteromalla				I		
Dicranum fuscescens						
Dicranum majus						
Dicranum scoparium				I		
Diplophyllum albicans						
Eurhynchium praelongum	II	III	III	I	IV	III
Eurhynchium striatum						
Fissidens taxifolius						
Hylocomium splendens						
Hypnum cupressiforme			I	I	I	I
Hypnum jutlandicum				I		
Isothecium myosuroides						
Leueobryum glaucum				I		
Lophocolea bidentata		I	I	I	I	I

W7	W8	W9	W10	W11	W12	W13	W14	W15	W16	W17	W18	W19
II	II	III	I	I			I					
II												
I	III	I	I		II		I					
I												
I	I	II	I									I
	I											
	I		I				I	III	I	I		
										I	I	II
				III						IV	III	II
		I	I	II			I	I	I	III	V	III
								I		II	I	
IV	IV	IV	II	II	II	I	I	I	I	I		I
I	II	IV	I	I								
	II	I			I							
		I		IV				I		IV	V	V
	I	II	I	II			I	II	I	II	I	III
									I	II	III	I
	I	I	I	II			I		I	II		
							II	I	I	I		
II	I	II	I	III				I		II	III	III

	W1	W2	W3	W4	W5	W6
Mnium hornum		II	IV	I	III	
Pellia epiphylla		I		II		II
Plagiochila asplenioides			I			
Plagiomnium affine			II			I
Plagiomnium rostratum			I		I	
Plagiomnium undulatum	I	I			II	I
Plagiothecium denticulatum		II		I	I	I
Plagiothecium undulatum						
Pleurozium schreberi						
Polytrichum formosum					I	
Ptilium crista-castrensis						
Rhizomnium punctatum		I	IV		II	
Rhytidiadelphus loreus						
Rhytidiadelphus squarrosus	I			I		
Rhytidiadelphus triquetrus						
Scapania gracilis						
Scleropodium (= Pseudoscleropodium) purum				I		
Sphagnum capillifolium				I		
Sphagnum fallax (= S. recurvum)		III	I	III		
Sphagnum fimbriatum		III		II	I	
Sphagnum palustre		III	I	II	I	
Sphagnum quinquefarium						
Sphagnum squarrosum		III	I	I	I	
Thuidium tamariscinum	I					

W7	W8	W9	W10	W11	W12	W13	W14	W15	W16	W17	W18	W19
III	II	III	II	II	I	I	III	III	I	III		II
I	I	I										
I	I	II	I							I		III
	I			I								
	I	I	I									II
III	III	IV	I	II	I							II
I	I	I	I	I		I		I	I			II
			I	I			I	I	I	IV	IV	III
I				III					I	IV	V	III
		I	I	II			I	II		V	I	II
										I	III	I
	I	I	I							I		
	I	I		I		I		I	I	IV	IV	III
		I		IV						II		III
	I	II		III					I	I	III	III
									I	II	I	
			I	IV					I	II	III	III
											III	
I												
										I		
I										I		
				I						I	III	
I												
II	I	IV	I	IV		I	I	I	I	III	I	IV

Appendix IV: The distribution of NVC data available for woodlands

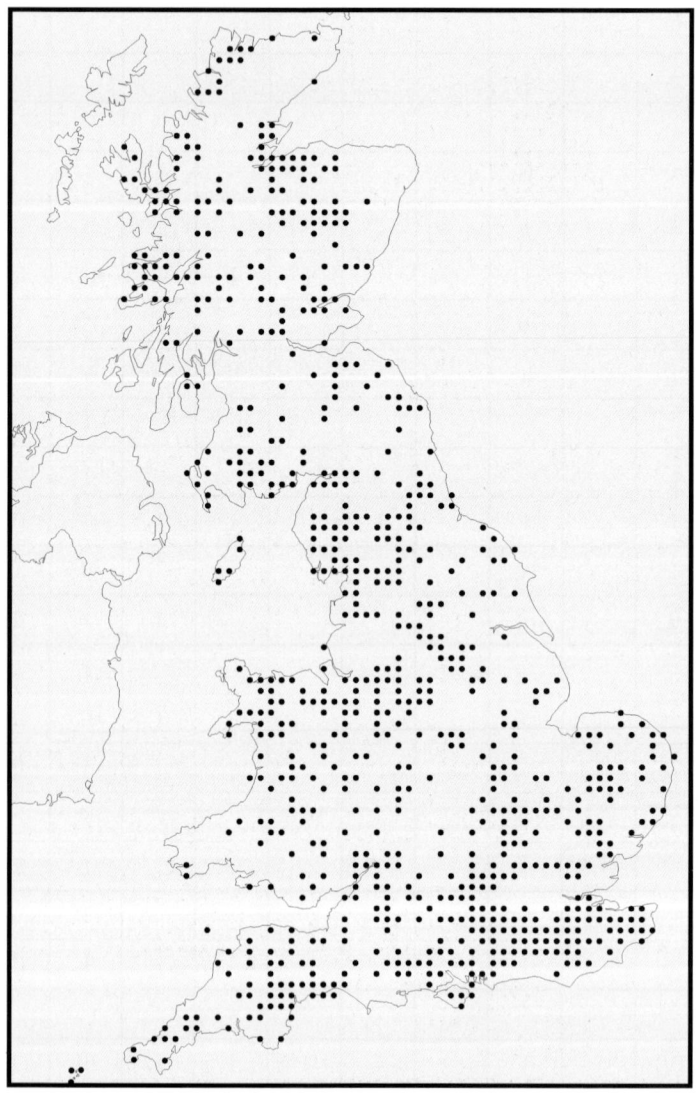

Figure 22 Available woodland community data as published in Rodwell (1991)

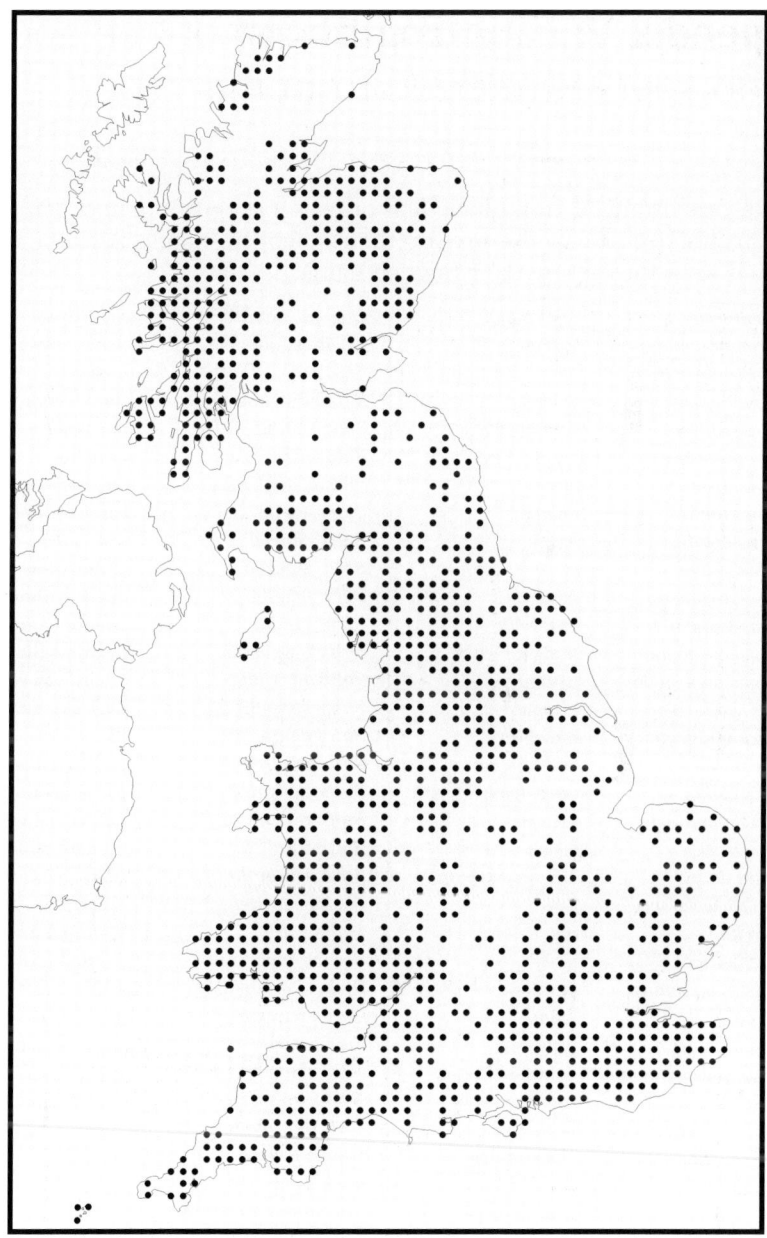

Figure 23 Available woodland community data held on English Nature's database (as at 01/01/2000)

Appendix V: Latin-English list of tree and shrub species referred to in this report

Species are arranged in alphabetical order by scientific name. Those in bold are referred to in the text by their scientific name. All others are referred to in the text by their English name.

Acer campestre	field maple	*Quercus petraea*	sessile oak
Acer pseudoplatanus	sycamore	*Quercus robur*	pedunculate oak
Aesculus hippocastanum	horse chestnut	*Rhamnus catharticus*	buckthorn
Alnus glutinosa	alder	*Rhododendron ponticum*	rhododendron
Betula pendula	silver birch	**Rosa canina**	dog rose
Betula pubescens	downy birch	**Rubus fruticosus**	bramble
Buxus sempervirens	box	**Ruscus aculentus**	butcher's-broom
Calluna vulgaris	heather	*Salix aurita*	eared willow
Carpinus betulus	hornbeam	*Salix caprea*	goat willow
Castanea sativa	sweet chestnut	*Salix cinerea*	grey willow
Cornus sanguinea	dogwood	*Salix fragilis*	crack willow
Corylus avellana	hazel	*Salix lanata*	woolly willow
Crataegus laevigata	Midland hawthorn	*Salix lapponum*	downy willow
Crataegus monogyna	hawthorn	*Salix myrsinites*	whortle-leaved willow
Cytisus scoparius	broom	*Salix myrsinifolia*	dark-leaved willow
Empetrum nigrum	crowberry	*Salix pentandra*	bay willow
Erica cinerea	bell heather	*Salix phylicifolia*	tea-leaved willow
Erica tetralix	cross-leaved heath	*Salix purpurea*	purple willow
Euonymus europaeus	spindle	*Salix reticulata*	net-leaved willow
Fagus sylvatica	beech	*Salix triandra*	almond willow
Frangula alnus	alder buckthorn	**Salix viminalis**	osier
Fraxinus excelsior	ash	*Sambucus nigra*	elder
Hedera helix	ivy	*Sorbus aria*	whitebeam
Ilex aquifolium	holly	*Sorbus aucuparia*	rowan
Juniperus communis	juniper	*Sorbus torminalis*	wild service
Larix spp.	larch	*Taxus baccata*	yew
Ligustrum vulgare	privet	*Tilia cordata*	small-leaved lime
Lonicera periclymenum	honeysuckle	*Tilia platyphyllos*	large-leaved lime
Malus sylvestris	crab apple	*Tilia x vulgaris*	common lime
Pinus sylvestris	Scots pine	**Ulex europaeus**	gorse
Populus nigra	black poplar	*Ulmus glabra*	wych elm
Populus tremula	aspen	*Ulmus* spp.	elm
Prunus avium	wild cherry	**Vaccinium myrtillus**	bilberry
Prunus padus	bird cherry	**Vaccinium vitis-idaea**	cowberry
Prunus spinosa	blackthorn	*Viburnum lantana*	wayfaring tree
Pseudotsuga menziesii	Douglas fir	*Viburnum opulus*	guelder rose

Appendix VI: A minimalist approach to data collection for use with the NVC key

The original samples were a mixture of fresh data collected to a standard protocol and samples collected previously in a variety of formats. The standard samples used a 50 x 50 m quadrat for tree and shrub data and either 4 x 4 m or 10 x 10 m quadrats for the ground flora (according to the nature of the vegetation). While quadrats were usually square the shape of the recording area was varied where this was necessary to avoid sampling a different homogenous stand (Rodwell 1991).

Subsequent experience has shown that it is possible to classify samples taken using a range of different quadrat sizes. In surveys carried out within NCC and English Nature the following approach has been adopted for collecting species data purely for NVC identification purposes in order to reduce the time required, particularly with respect to the tree and shrub layer.

(a) Identify the area to be sampled, the 'homogenous stand', by a quick walk over the area.

(b) Lay out a 5 x 5 m ground flora plot at the first sampling point. (5 x 5 m was settled on rather than 4 x 4 m because it fitted in better with a wide range of previously collected data.)

(c) Record species presence and Domin cover for each species in the ground flora plot.

(d) Record tree and shrub species presence and cover over the plot and in the immediate surroundings, i.e. within 10-15 m all round of the plot. The area sampled is not as big as the original 50 x 50 m samples, but experience suggests that tree and shrub species that are going to be significant in the classification process are unlikely to be so infrequent that they will be missed by this process. The advantage is a significant reduction in the time taken.

(e) Move to another area of the stand and lay out a second ground flora plot. Normally five plots should be recorded unless the stand is very small.[1] They should be distributed evenly over the stand area.

---

1. Note that constructing a constancy table from five plots from the same stand is not comparable with the methodology used to construct the tables published in *British Plant Communities*, which summarise between stand variation. The over-description of local peculiarities should therefore be considered when making community determinations.

Note that this approach is intended simply for the taking of samples for NVC identification purposes. It is not meant as a protocol for quadrat sampling for other woodland survey purposes or for monitoring. These may require different sizes or distributions of plots according to the purposes of the survey.

(f) Repeat the whole recording process (a-d).

A simple form is reproduced below, which can be used to construct a frequency table in the field.

Site _____ **Recorder** _____
Compartment _____ **Date** _____
Brief description _____

_____

_____

Sample (5 x 5 m plot plus surrounding canopy)	1	2	3	4	5	Freq.	Domin range
Tree layer species							
Shrub layer species							

Domin 1: rare, 2: sparse 3: frequent (but all less than 4%), 4: 4-10%, 5: 11-25%, 6: 26-33%, 7: 34-50%, 8: 51-75%, 8: 51-75%, 9: 76-90%, 10: 91-100%

Sample (5 x 5 m plot)	1	2	3	4	5	Freq.	Domin range
Ground flora species							

Domin 1: rare, 2: sparse, 3: frequent (but all less than 4%), 4: 4-10%, 5: 11-25%, 6: 26-33%, 7: 34-50%, 8: 51-75%, 9: 76-90%, 10: 91-100%